P-51 MUSTANG WARBIRDS

NICHOLAS A. VERONICO, JIM DUNN, AND DAVID LEININGER

Featuring the photography of:
Roger Cain, Keith Charlot, Luis Drummond, Jerry Liang, Milo Peltzer, Jim Raeder, Taigh Ramey, Mike Shreeve, and Richard H. VanderMeulen

Key Books

Front cover: Sold surplus in 1958, this Mustang was acquired by Trans Florida Aviation and converted to Cavalier 750 executive configuration. From 1968 to 1974, the El Salvadorean Air Force flew it in Latin American skies. Well-known Mustang operator Vlado Lenoch owned and flew this Mustang at numerous airshows to the delight of spectators across the country. The fighter is now owned by Paul Wood's Moonbeam Historic Military Aircraft of Waukegan, Illinois. (David Leininger)

Back cover: P-51D-30-NT 45-11471 spent the majority of its military career serving in a number of Air National Guard units. Sold surplus at the May 1958 auction at McClellan Air Force Base (AFB), it flew as a civil Mustang until it crashed in 1979. Rebuilt as the scoop-less air racer Stiletto, it was campaigned around the Reno pylons by Skip Holm, as well as Scott and Denny Sherman. After its racing career was over, the aircraft was rebuilt to stock configuration as a dual-control TF-51D named *Diamondback*. (Roger Cain)

Title page image: P-51D-30-NA 44-75009 *Rosalie* was sold surplus at one of the auctions at McClellan AFB, California, in 1958. The Mustang was in the family of David and Homer Roundtree from 1965 to 1985, when it was sold to Ted Contri. In 1988, *Rosalie* joined Contri's other Mustang *Sizzlin' Liz* in the family stable. Contri also owned 44-84753, which today flies as *Buzzin Cuzzin*. (David Leininger)

Contents page image: Thomas T. Ungurean of Coshocton, Ohio, acquired P-51D-30-NA 44-74813 in 2014. The aircraft wears the colors of Maj. George E. Preddy, Jr.'s wartime Mustang 44-14906 *Cripes A'Mighty*, which flew with the 352nd Fighter Group based at RAF Bodney. Preddy was credited with 26.833 aerial victories. He perished when his Mustang was shot down by Allied anti-aircraft fire on Christmas Day, December 25, 1944. (Richard H. VanderMeulen)

Published by Key Books
An imprint of Key Publishing Ltd
PO Box 100
Stamford
Lincs PE19 1XQ

www.keypublishing.com

The rights of Nicholas A. Veronico, Jim Dunn, and David Leininger to be identified as the authors of this book has been asserted in accordance with the Copyright, Designs and Patents Act 1988 Sections 77 and 78.

Copyright © Nicholas A. Veronico, Jim Dunn, and David Leininger, 2022

ISBN 978 1 80282 301 1

All rights reserved. Reproduction in whole or in part in any form whatsoever or by any means is strictly prohibited without the prior permission of the Publisher.

Typeset by SJmagic DESIGN SERVICES, India.

CONTENTS

Introduction — 4

Chapter 1 – Selling Surplus Mustangs — 5

Chapter 2 – Warbird Mustangs in Their Element — 27

Chapter 3 – Mustang Gate Guardians: In from the Cold — 98

INTRODUCTION

North American Aviation's P-51 Mustang evolved into the ultimate long-range escort and air superiority fighter of World War Two. The combination of the Mustang's laminar flow airfoil and the Rolls-Royce V-1650 Merlin engine gave the Mustang the range and performance needed to combat the Luftwaffe in the skies over Europe. In addition to the United States and England, the air arms of 29 other nations flew the Mustang, the last being the Dominican Republic, which retired its P-51s in 1984.

As the Mustang was retired from service, surplus P-51s[1] became available to the civilian market. In the early 1960s, they became fast executive transports and high-performance sport planes, while some were preserved as gate guardians at various military bases and Air National Guard (ANG) fields around the country. Approximately 35 years after the end of the war, the first of the civil Mustangs was painted in World War Two markings, giving fuel to the then-growing warbird movement. Sixty years after the end of World War Two, the quest for authenticity shifted into high gear. No longer was it good enough to restore an aircraft, it has now become the ultimate drive to return Mustangs to their factory-fresh appearance. By 2022, this quest for authenticity has spread throughout the warbird and wider aviation community, where aircraft from antiques and classics to L-birds and heavy bombers are going back to original condition.

According to the website MustangsMustangs.net, there are 299 surviving P-51s of all types, with 174 flying aircraft, 59 on display, 45 in long-term maintenance or under restoration, 21 in storage, with nine unaccounted for. The numbers of flying aircraft and those in maintenance or restoration will fluctuate over the next few years, as eight to ten aircraft will soon be returning to the sky. As each new restoration flies, it moves the standards for authenticity to increasingly higher levels.

Airshows each year give enthusiasts the opportunity to see Mustangs in the air and hear the Merlin engines' growl. For those fortunate enough, living history flight experiences are the ultimate opportunity to experience the P-51's performance from the cockpit. Dual control Mustangs are no longer a rarity and flight instruction is now an option, one that has improved the safety and training of P-51 pilots both new and experienced.

Mustang Warbirds is a visual tribute to the engine builders, aircraft restorers, maintainers, owners, and pilots who devote so much time and attention to preserving North American's P-51 Mustang.

---

1. Note that the P-51D was redesigned from P-Pursuit to F-Fighter by the US Air Force (USAF) in 1948. For consistency, P-51 will be used interchangeably to mean both P-51 and F-51 throughout.

CHAPTER 1
SELLING SURPLUS MUSTANGS

In 1944, as the outcome of World War Two ending with an Allied victory was coming into focus, policymakers in the United States started to address the issue of surplus aircraft. In the years following World War One, the United States was flooded with inexpensive surplus Curtiss Jennies and de Havilland DH-4s, along with thousands upon thousands (11,900 to be exact) of new, in-the-crate, 12-cylinder Liberty engines. This massive surplus stymied the American aviation industry well into the 1930s, putting the US military far behind other nations in terms of aeronautical innovation and engine development. To illustrate the point, the National Guard bureau was still flying Curtiss JN-4 Jennies in 1927, nearly a decade after the end of World War One. In comparison, Britain's Royal Air Force (RAF) introduced the Armstrong Whitworth Siskin IIIA, the service's first all-metal fighter, that same year.

The task of studying the postwar problem of surplus aircraft was given to The Harvard University School of Business Administration in early 1944. The product of this study, known as the *Harvard Report*, made recommendations for which types of aircraft should be offered to the civilian market, to commercial air carriers, as well as the disposition of tactical aircraft. An "operational reserve" of aircraft was to be set aside for future use by the US Army Air Forces (USAAF), which included P-51s and P-47s, C-47s, B-25s and B-29s. Everything else was to be sold or recycled; the recycled aluminum would then feed consumer demand for goods, such as home appliances and automobiles, that had been unavailable during the war. Strategic aircraft reserves were set up in the United States at fields including Davis-Monthan Army Air Field, Arizona, and Pyote Army Air Field, Texas, as well as in England and Germany.

Victory over the Nazis came on May 7, 1945, and the Japanese surrender followed four months later on September 2, 1945. Those two events grounded thousands of aircraft and ended production contracts for thousands of additional USAAF and Navy planes.

By June 1946, more than 27,000 tactical aircraft were declared surplus. Fields of aircraft were scrapped in Europe and on islands in the Pacific Theater, as well as within the United States. On June 10, 1946, the War Assets Administration requested bids to scrap 23,663 tactical aircraft stored at Kingman, Arizona; Walnut Ridge, Arkansas; Ontario (Cal-Aero Field), California; Clinton, Oklahoma; and Albuquerque, New Mexico. The winners of the five fields in question, were, essentially, construction companies, and they formed a partnership to determine how to extract the most profit from the aircraft. The partnership was called the Aircraft Conversion Corp., and the managing partners were George R. Brown and Herman Brown – great builders whose company became Brown and Root, later Kellogg Brown and Root, at one time part of Halliburton. They sold aircraft parts back to the government, hired a metallurgist to design a furnace for reducing the aircraft to ingots, and built the same model furnace at each field. The metallurgist also determined the aircrafts' smelting order to produce the cleanest metal alloy, thus yielding the most profit.

As the Flying Fortresses, Liberators, Lightnings, Airacobras, Kingcobras, and Warhawks were fed into the furnaces, what was missing were Mustangs. The majority of the P-51A, B, C, and A-36A Mustang production had been shipped overseas, while only a small percentage was retained to meet the pilot and mechanic training needs stateside, and many of those aircraft were lost in training accidents. P-51Ds in theater were retained as a reserve or returned to the United States if they were new or low-time. War-weary Mustangs were scrapped in theater. Mustangs within the continental United States that were new formed part of the USAAF's operational reserve, while some training aircraft were sold surplus.

Aviator Paul Mantz and his investors purchased the entire field of surplus aircraft stored at Stillwater, Oklahoma. Of the 475 planes Mantz acquired, there were only seven P-51Cs and a single A-36A in the lot. The same lack of Mustangs was seen at Walnut Ridge, and there was only one Mustang among the 5,463 aircraft at Kingman. Thus, very few P-51s made their way to the post-World War Two civil market, and those that did were stripped for use at the National Air Races at Cleveland, Ohio.

The end of World War Two saw thousands of training and tactical aircraft become surplus to the needs of the US Army Air Forces (USAAF). Although P-51s were offered for sale, they were very few in numbers. Mustangs at combat bases overseas were typically scrapped in theater, while new-build and newly delivered Mustangs were held to form new, postwar squadrons and to form a strategic reserve of fighter aircraft. The B-25 and stateside trainer P-51D are seen at Reconstruction Finance Corp. (RFC) storage depot at Stillwater, Oklahoma. In 1946, famed aviator Paul Mantz led a consortium of buyers who bought the entire field of 478 aircraft. (Oklahoma State University via Woody Harris/Scott Thompson)

Some of the very few high-back Mustangs available for sale were located at the RFC Ontario depot, which is today's Chino Airport in southern California. The RFC named their depots after the largest municipal locality, or, in the case of surplus aircraft, the nearest town with an air base. During the war years, Ontario Army Air Field was home to the 364th and 329th Fighter Squadrons as well as the 443rd Base Unit (Combat Crew Training Station – Fighter). A-36A 42-83736 is seen at RFC Ontario (Chino Airport) in 1946. This Mustang did not find a buyer and was scrapped. (Emil Strasser via Jerry Liang)

Sources of Civil Mustangs

During World War Two, the Royal Australian Air Force (RAAF) received 100 P-51Ds in kit form to be assembled by the Commonwealth Aircraft Corp. (CAC) at Melbourne, Victoria. Eighty of the kits were assembled and designated CA-17 Mustang Mark 20. In addition, 84 P-51Ks were sent to Australia by ship. After the war, in 1946, CAC built 120 additional P-51Ds designated CA-18 Mustang Mark 21 (V-1710-7 engine), Mark 22 (reconnaissance equipped), or Mark 23 (Merlin 66 or 70 engine) depending upon configuration. When the RAAF phased the type out of service in 1959, dozens of Mustangs were put up for sale.

Back in 1953, six Mustangs were used as targets in Australia's Emu Field nuclear tests. Two blasts, known as Operation *Totem One* and *Totem Two*, took place in October 1953, using devices triggered on towers. After the blasts, the Mustangs sat on the ranges until the mid-1960s, when the Australian government put them up for disposal. Warbird historian Geoff Goodall and a party of friends were able to inspect and photograph the aircraft in 1967. All six of the Mustangs were sold to civilian buyers and were subsequently recovered from the test range. All became warbird projects, and today RAAF serial A68-1 is listed as a flying example in the United States.

While the mainstay of the postwar USAAF's and US Air Force's (USAF) propeller-driven fighter force was the P-51D and P-51H, many surplus Mustangs were sold to foreign air arms through various military defense aid programs. The Royal Canadian Air Force (RCAF) purchased 150 Mustangs in the postwar years. When the RCAF Mustangs were sold off, two went to museums and 87 found civilian buyers. Of those, 71 were purchased by Intercontinental Airways of Canastota, New York, which was an investor group headed by James Defuria. Having acquired the Mustangs in late 1956 through 1958, Defuria's partnership in turn sold dozens into the US civil market, as well as a number to the air forces of El Salvador and Guatemala.

During the war, ten damaged Mustangs sought neutral Sweden as a safe haven rather than crash landing in Nazi-held territory. In addition to the interned Mustangs, Sweden acquired 43 P-51s in April 1945, as the war was reaching its final climax. This batch of Mustangs set the Swedes back US$160,000 for each aircraft. Ninety additional Mustangs were bought after the war in 1946, and because of the glut of surplus P-51s, only cost US$3,500 each. In 1948, Sweden ordered an additional 21 Mustangs. The type was flown into the early 1950s, by which time attrition had claimed 72 in crashes or they had been striped for parts. In June 1952, the Israeli Defense Force/Air Force (IDF/AF) bought 25 Mustangs from Sweden. That sale was followed by 42 sold to the Dominican Republic in the winter of 1952/53, and 25 to Nicaragua later in 1953.

Switzerland added the Mustang to its air force in 1948, when the always-neutral nation purchased 130 from surplus US stocks. The type was flown for a decade and was retired in April 1958. Nearly all of the Swiss Mustangs were consumed as targets or simply scrapped. Some parts from the gunnery ranges did escape to be incorporated into other Mustang projects, but not many.

The Italians received 173 of the type under the Military Defense Assistance Program between 1947 and 1951. In 1955, the Italians sold 30 Mustangs to Israel, and continued to fly the type until 1958. That year, the Italians gave eight aircraft to the Somalian Air Force, while the balance of the fleet was scrapped. One P-51D, Italian Air Force serial number MM4324, ex-44-73451, is displayed at the Italian Air Force

In the postwar years, P-51 Mustangs formed a strategic reserve of fighter aircraft and would be called up to equip numerous Air National Guard (ANG) units and to support Allied troops during the Korean War. Notice that the canopies have been opened approximately eight inches, and a wooden venting cover has been installed to maintain a constant interior temperature. (National Archives via Kevin Grantham)

Museum at Vigna di Valle, near Rome, and a second, MM 4309, was found in Lake Guarda, Italy, in 2013. MM 4309 was lost on August 7, 1951, along with pilot Lt. Paolo Tito, who perished in the crash. This substantial wreck was recovered in 2016.

The Royal New Zealand Air Force (RNZAF) was slated to receive 370 Mustangs in mid-1945, however, the war ended and only 30 were delivered. These were stored until 1951. Their service life was quite short, as the type was retired in 1955 and subsequently sold off in 1958. One example preserved by the late John Smith, P-51D-30-NT 45-11513 (RNZAF serial NZ2423), was sold following Smith's passing and is under restoration by Brenden Deere, at RNZAF Ohakea, on the North Island. Seven aircraft and many other parts worked their way into the civilian Mustang fleet.

McClellan Auctions, Cavalier, and Mustang Recoveries

Once the USAF and ANG made the decision to discontinue operating the Mustang, the Air Material Command was tasked with disposing of the aircraft and the associated parts supply. It was decided to sell the Mustangs at auction, which was held at McClellan Air Force Base (AFB), outside Sacramento, California. The first of the McClellan auctions opened on September 3, 1957, and offered 75 aircraft in a sealed-bid sale. Only 38 of the 75 Mustangs sold, all to civilian buyers. Two additional auctions were held, with nearly all 75 P-51Ds selling.

Simultaneous to the US government auctions at McClellan AFB in late 1957 and 1958, other air arms were divesting themselves of the Mustang in favor of new jet fighters. When Mustangs were auctioned to the public, David B. Lindsay, Jr., publisher of the Sarasota, Florida, *Herald-Tribune* newspaper and president of Trans Florida Aviation, saw an opportunity to provide fast, executive transport aircraft, and he recognized there was a market for military-configured Mustangs in Central and South America. He called his company Trans Florida Aviation, the name of which was later changed to the Cavalier Aircraft Corp. The executive conversions featured a tall tail, a second seat, and the Cavalier Model 2500 was fitted with wingtip fuel tanks. As the LearJet eclipsed surplus Mustangs as an executive transport, Lindsay's business focused on P-51D conversions for military use. Cavalier's military Mustangs featured strengthened wings with additional hard points, a second seat, transport versions of the Merlin engine, and increased fuel capacity/range.

Cavalier Aircraft Corp. first landed a contract to inspect and repair as necessary (IRAN) three dozen Mustangs for the Dominican Air Force. These aircraft also received a number of Cavalier upgrades. A contract for nine modified Mustangs (seven P-51Ds and two TF-51Ds) came in 1967, and these were delivered under the US Department of Defense's Peace Condor program to Bolivia. In 1968, four Cavalier Mustang IIs and one TF-51D were built for El Salvador. The El Salvadorean Mustangs would soon see the Mustang's final combat during the 100 Hour War with Honduras (July 14–18, 1969). This was the last hostile engagement for the P-51 and the last air-to-air combat by reciprocating engine fighters when the El Salvadorean Cavalier Mustangs engaged Honduran F4U Corsairs.

A number of the Swedish Mustangs that had been sold to Nicaragua ended up in El Salvador. The late W. W. "Will" Martin began ferrying the planes back from Nicaragua in 1963. The following year, Martin acquired Costa Rica's two remaining Mustangs (44-73193 and 44-74978). During the ferry flight back to the United States, 44-73193 suffered an engine failure and was damaged beyond economical repair in the off-field landing. Martin chronicled his adventures ferrying home the Mustangs in his 2013 book *So I Bought an Air Force: The True Story of a Gritty Midwesterner in Somoza's Nicaragua*.

P-51C-5-NT 42-103725 spent its USAAF career on the US East Coast and was stored at RFC Stillwater at war's end. To get around the restriction on selling tactical aircraft, Paul Mantz's partners, L. B. Hapgood and J. W. Heath, formed H&H Enterprises in Henrietta, Texas. H&H Enterprises sold "Mustang kits," and, in this case, a kit was sold to Frank J. Abel of Wichita Falls, Texas. Abel had assembled a number of kits and NX4814N was just one of many. The fully assembled aircraft was sold to Jack Hardwick in July 1947. It is seen wearing the name *Batty Betty* at the National Air Races in Cleveland, Ohio, just days before Hardwick crashed during the races. Rumor has it that although Hardwick crashed, he was the race's big winner. He had heavily insured *Batty Betty*, and the story says that when firefighters came to extinguish the fire, Hardwick waved them off in an effort to let the fighter burn completely so he could collect the insurance money. (Emil Strasser via Jerry Liang)

Seen at summer camp in Boise, Idaho, in 1952, California ANG P-51Hs taxi in upon arrival. The lightweight P-51H Mustangs were kept stateside to equip ANG units while P-51Ds were sent to Korea to perform the fighter/bomber role in support of Allied troops. (Al Hamblin)

Seen in June 1967, Commonwealth Aircraft Corp. (CAC) Mustang A68-1, the first built by CAC in 1945, rests at the Australian nuclear test site at Emu. Six CAC Mustangs (A68-1, -7, -30, -35, -72, -87) were part of the tests, and all made their way to the US civil market. The Royal Australian Air Force (RAAF) received 215 P-51Ds and 84 P-51Ks under the lend-lease program, while CAC built another 200 (80 from kits supplied by North American Aviation). Australia became a huge source of Mustangs and Mustang parts in the late 1960s and early '70s. (Geoff Goodall)

P-51D-30-NA 44-74376 was in the batch of 32 Mustangs transferred to Canada in June 1947 and is seen on the seventh of the month upon arrival. The P-51D wears the Royal Canadian Air Force (RCAF) roundel and fin flash but retains its USAAF serial number. The RCAF assigned this aircraft serial number RCAF 9578. The RCAF phased out its Mustangs in 1960, and many wound up flying in Central and South American countries. (Nicholas A. Veronico Collection)

P-51D-20-NA 44-63992 was accepted by the USAAF on December 19, 1944, and shipped to Great Britain, departing on March 5, 1945. Stored in England for the duration of the war and months after, this aircraft was part of a 90 Mustang order for the Swedish Air Force. The aircraft were designated J-26s, and this aircraft was given serial number 26020. Sweden operated the type until 1953, when the remaining aircraft in the fleet were sold off to Israel, the Dominican Republic, and Nicaragua. Israel donated this aircraft to the Swedish Air Force Museum at Malmslätt. (William T. Larkins Collection)

The Swiss Air Force purchased 130 P-51D-20 and -25s that were being held in reserve in Germany in 1948. The Swiss paid US$4,000 for each aircraft. Swiss P-51D-25-NA serial number J-2113 (ex 44-73349) is the sole remaining example of the type in the country, and it resides at the Swiss Air Force Museum at Zürich's Dübendorf Airfield. (Roland Poehlmann)

In subsequently years, a few American owners sold their Mustangs south of the border, and they ended up in Bolivia in 1966 and El Salvador in 1969. By 1974, businessman Jack Flaherty of Monterey, California, had recovered nine Mustangs from El Salvador, including one rare TF-51D conversion.

In 1972, five Cavalier Mustang IIs and a single TF-51D were delivered to the Indonesia Air Force (Tentara Nasional Indonesia Angkatan Udara – TNI-AU), which already had a sizable Mustang force. In total, the TNI-AU operated 49 Mustangs. Ralph Johnson, his son Stephen Johnson, and Art Stagg, all business partners from northern California, recovered 15 Mustangs along with a substantial amount of spare parts beginning in 1972.

The Uruguayan Air Force (Fuerza Aérea Uruguaya, FAU) had acquired 25 Mustangs during the 1950s, and they subsequently sold to five to Bolivia. Canadian Arny Carnegie led the effort to return seven Mustangs from Bolivia in 1972. They were flown out as a group. In total, 18 P-51s are known to have been recovered from Bolivia and returned to the warbird market.

Left: In 1957 and 1958, the USAF auctioned more than 100 Mustangs of various models. The bidding was managed through Norton AFB in southern California, while the aircraft were stored at McClellan AFB, outside Sacramento, California. This ad from the December 16, 1957, issue of *Aviation Week* offers 17 F-51Ds on auction, which was slated to close on January 10, 1958. (Nicholas A. Veronico Collection)

Opposite: P-51D 45-11553 sits at McClellan AFB's Mat I storage area in 1957. Note the other P-51Ds up for auction parked to the left rear. Mustang 45-1153 was being sold with 1,439.2 hours total time airframe and 468.4 hours on the engine. A number of Texas ANG Mustangs, including this aircraft, participated in the filming of the Rock Hudson movie *Battle Hymn*. R. D. Nesen of Oxnard, California, paid US$879.79 for the Mustang, which he registered as N5414V. Today, N5414V flies as N51VF *Shangri-La*. (Milo Peltzer Collection)

Left: David B. Lindsay, Jr. (1922–2009) was the publisher of Sarasota's *Herald-Tribune* newspaper and visionary founder of Trans Florida Aviation, later changed to Cavalier Aircraft Corp. Lindsay was quick to recognize the profit potential of refurbishing P-51 Mustangs for the executive transport and military markets. His business sense saw Cavalier purchase the FAA Type Certificate (TC-11) from North American Aviation, enabling it to build virtually any Mustang part for its Cavalier conversions. Years later, Cal Pacific Airmotive of Salinas, California, acquired the Type Certificate, enabling it to supply Mustang parts to the warbird community. (Edward Lindsay Collection)

Opposite: Ground crews undergo training on a new Cavalier Mustang II, exported to a Latin American country under the US Military Assistance Program (MAP). The Mustang II featured a strengthened wing with up to six hardpoints under each wing, wingtip fuel tanks, a second seat for an observer, as well as a V-1650-724 transport engine. (Edward Lindsay Collection)

Selling Surplus Mustangs

Israel's 25 ex-Swedish Mustangs were added to the others they had acquired for an eventual fleet of 79 operational Mustangs with 26 airframes for a parts source. As the Israeli Air Force transitioned to jet fighters, including the Gloster Meteor and the Dassault Mystère and Ouragan, only one squadron of Mustangs remained active, owing to the piston-powered fighter's range over the new, short-legged jets. In 1976, collector Robs Lamplough recovered four IDF/AF Mustangs, which started a pilgrimage of sorts for other restorers, most notably brothers Angelo and Peter Regina. Peter Regina found a P-51B main aircraft and, using an Israeli Mustang fuselage for the basis, built the P-51B *Shangri-La*, today's *Princess Elizabeth*.

The Guatemalan Air Force acquired 30 Mustangs through various programs beginning in summer 1954. Don Hull was able to acquire the surplus Guatemalan Mustangs along with that air force's cache of spare parts in August 1972. The following year, Hull sold his collection to Wilson C. "Connie" Edwards, who stashed them away in a hangar on his property until the late 1990s to 2000, when he began to sell off some of his collection.

Opposite: P-51D-25-NT 44-84745 was sold surplus at the September 1957 McClellan auction for US$1,053 to Joe K. Hammer of Sacramento, California. The aircraft went to Cavalier Aircraft and was stored until June 1984, when acquired by Gordon Plaskett of King City, California, who rebuilt it into TF-51 configuration using parts from two other Mustangs. This aircraft now flies as N851D, the second Mustang to wear that registration, with Lee Lauderback's Stallion 51 Corp. in Kissimmee, Florida. (Milo Peltzer Collection)

Right: TF-51s *Crazy Horse* (44-84745, N851D) and *Crazy Horse²* (44-74502A, NL351DT) are owned and operated by Lee Lauderback's Stallion 51. Both aircraft wear the paint scheme of Mustangs flown with the 487th Fighter Squadron, 352nd Fighter Group, Eighth Air Force, when stationed at Asche, Belgium. Stallion 51 has been a leader in P-51 transition training and checkouts, recurrent flight training, formation flying, and orientation flights. (Richard H. VanderMeulen)

Right: Seen at Willows Airport, north of McClellan AFB, after being removed from storage, Ed Maloney, perched on the cockpit rail, was able to get this Dallas-built P-51D-30 donated to The Air Museum, then located at Claremont, California. Registered N5441V, serial number 45-11582, this is most likely the only Mustang still in possession by its original post-auction owner. (Milo Peltzer Collection)

Opposite: Maloney used some type of removable paint, as 45-11582 had to leave McClellan AFB without military markings. When the Mustang arrived in Claremont, the temporary paint was removed to show its original, albeit a bit worn, West Virginia ANG markings. (Milo Peltzer)

The Dominican Air Force operated 52 varying types of the P-51A, C, D, F-6K, and Cavalier Mustangs. The first were acquired on the US civil market in 1948, while later aircraft came from the Swedish Air Force in 1952. Brian O'Farrell and his partner Armin Mattli struck a deal with the Dominicans in May 1984 to recover all flyable Mustangs and all spare parts. At least nine ex-Dominican Air Force Mustangs that passed through the O'Farrell–Mattli partnership have been restored to flying condition.

Dozens of Mustangs had been scattered across the globe to support various air forces. Once they were returned to their country of birth, it was time to put them back into the air. That is what the warbird movement is all about – seeing flying examples of combat aircraft, from both sides of the conflict, in their element.

Sold surplus at one of the McClellan auctions and registered N5480V, this Mustang passed through a number of civilian owners until 1966. It is seen that same year at Long Beach, California, before it was acquired by Stan's Aircraft Sales of Fresno. Stan's Aircraft had sold a number of Mustangs south of the border, and 44-73129 was involved in a deal with the aircraft broker. It is believed that 44-73129 was planned to be exported to Haiti but was actually diverted to El Salvador. It was recovered in 1972 and flies today as N151SE *Merlin's Magic*. (Milo Peltzer)

CHAPTER 2
WARBIRD MUSTANGS IN THEIR ELEMENT

In the more than 80 years since the first flight of the prototype North American Aviation P-51 Mustang, there have been five distinct eras – World War Two and Korea; ANG and foreign postwar service; Recoveries in the 1960s and '70s; the Warbird Movement; and the Ultra Authenticity era.

Each has served a purpose that has helped define and establish the P-51 Mustang legend. Of course, World War Two and Korea are where the aircraft saw action and made heroes out of the men and women who built the Mustang and those who faced the life-and-death duel in the sky with the enemy. The ANG and foreign postwar service era saw many states flying the Mustang in national defense, while more than one dozen foreign air forces used the P-51 to defend their sovereign nations from outside aggression. The Recovery period came about as the demand for Mustangs was increasing as an executive transport and the Warbird Movement was picking up speed. The Warbird Movement of the 1980s and 1990s saw many aircraft returned to airworthy status, and Mustangs in civil paint schemes were replaced by P-51s sporting military colors. The Ultra Authenticity era, where we are today, has seen the restorer's craft taken to an entirely new level.

The Ultra Authenticity era has seen aircraft returned to the condition they were in when each rolled off the assembly line at Inglewood or Dallas, or how they appeared in combat. Interior markings, those found on the assembly line, are skillfully recreated, cockpits and armament are fitted out to the smallest detail, while some restorations have gone to extreme lengths by doing things such as recreating the 108-gallon "paper" drop tanks, and so much more.

Safety of flight is another big consideration in today's era of Mustang operations. There is a tipping point in a restoration where inspect and repair as necessary is dropped and an aircraft has to come completely apart and be rebuilt to ensure that material failure does not take an aircraft and pilot out. Advances in computer technology have enabled parts thought to be extinct to be reproduced. Pilot proficiency training standards have been set, not by a government, but by peers in the Mustang and warbird community. The proliferation of two-seat, dual-control Mustangs, and those who instruct in them, have set a new standard for P-51 proficiency. After all, one does not own or fly a Mustang – they are caretakers of aviation history.

A Mustang must be seen in its element to be truly appreciated.

XP-51 41-038

Built: Inglewood, California

Service

1941: It was sent to US Army Air Forces, Wright Field, Ohio, for evaluation.
1942: It moved to NACA Langley, Virginia.
1945: It was in storage at Freeman Field, Indiana.
1949: It was in storage at the National Air & Space Museum, Washington, DC.
1975: It was then traded to the Experimental Aircraft Association.
Current Owner: Experimental Aircraft Association (NX51NA)
Based: Oshkosh, Wisconsin

This aircraft is one of two prototypes ordered by the USAAF. This aircraft was the fourth Mustang produced and the first delivered to the USAAF. The fighter was extensively tested by the USAAF and, in 1942, it was assigned to the National Advisory Committee for Aeronautics (NACA) and wind tunnel tested at NACA's Langley, Virginia, facility. After the war, it was stored for an anticipated US Air Force Museum. In 1948, the Air Force transferred its museum aircraft to the National Air & Space Museum, which held the fighter in storage until 1975, when the Experimental Aircraft Association (EAA) traded Northrop Alpha NC11Y for the XP-51. The fighter was rebuilt by Darrell Skurich in Colorado and was flown for six years, until being retired to the EAA Museum in Oshkosh, Wisconsin.

Opposite: The XP-51 taxies in after a flight demonstration at Harlingen, Texas, in October 1978. (Photo by John Kerr via Mark Hrutkay Collection)

Right: XP-51 NX51NA on the grass at EAA AirVenture 1982. (Photo by John Kerr via Mark Hrutkay Collection)

Left: XP-51 41-038 in storage at Freeman Field, Indiana, after the war. (Mark Nankivil Collection)

A-36A-1-NA 42-83731

Built: Inglewood, California; December 1942

Service

1942: It was delivered to the USAAF on December 16, 1942, at Long Beach, California.

1943–44: It served at Meridian Army Air Field, Mississippi; Harding Army Air Field, Louisiana; Mabry Army Air Field, Florida; Napier Field, Alabama; Lawson Army Air Field, Georgia; Kelly Field, Texas.

1945: It then went into storage at Albuquerque Army Air Field, New Mexico.

1947 (date estimated): The Mustang was purchased from the Reconstruction Finance Corp. by Jack Hardwick, and stored at El Monte, California.

1977: It was sold to Thomas Camp, Livermore, California, as a project aircraft.

1978: The A-36 was acquired by Dick Martin and registered as N50452.

1980: It was sold to Friedkin Enterprises, Houston, Texas, and reregistered as N251A.

2012: It was transferred to Comanche Warbirds, Inc., Houston, Texas.

Current Owner: Comanche Warbirds, Inc., Houston, Texas (N251A)

Current Scheme Honors:
Aircraft from the 86th Fighter-Bomber Group, which flew from bases in Sicily, Italy, Corsica, France, and Germany.

Opposite: Comanche Warbirds' A-36 in formation with the collection's P-51C *Princess Elizabeth*. (David Leininger)

Right: The beautiful lines of the A-36 Mustang, showing the two 0.50-cal. machine guns in the lower cowling and a pair in each wing, with a pair of 500lb (226.8 kg) bombs on the underwing hardpoints. (David Leininger)

Left: After the war, 42-83731 was stored at the RFC's depot at Albuquerque, New Mexico. The sleek fighter was acquired by Jack Hardwick and moved to his facility in El Monte, California, for storage. (Milo Peltzer Collection)

A-36A-1-NA 43-83738

Built: Inglewood, California; December 1942

Service

1942: It was delivered to the USAAF on December 21, 1942, at Long Beach, California.

1943–44: It served at Memphis, Tennessee; Key Field, Mississippi; Waycross, Georgia; Will Rogers Field, Oklahoma; New Orleans and Harding, Louisiana; DeRidder Army Air Field, Louisiana; Kelly Field, Texas.

1945: On June 25, the Mustang was transferred to storage with the RFC.

1946: It was sold to Douglas Aircraft Co., Santa Monica, California, and registered as N4607V.

December 1948 to March 1980: The Mustang passed through various owners in the United States.

1980–2014: It was acquired by Max R. Hoffman, Ft. Collins, Colorado; then sold onto John R. Paul, Hamilton, Montana, in 2009. It was then acquired by the Collings Foundation, Stowe, Massachusetts, in 2014.

Current Owner: Collings Foundation, Stowe, Massachusetts (N4607V)

Current Scheme Honors:

A-36A *Baby Carmen*, 526th Fighter-Bomber Squadron, 86th Fighter Group, Mediterranean Theater of Operations.

Opposite: Up close, you can see the dive-brakes deployed on A-36 *Baby Carmen*, seen here shortly after restoration. (David Leininger)

Right: *Baby Carmen* was restored by American Aero Services in New Smyrna Beach, Florida. (David Leininger)

Left: In the 1960s, Sid Smith owned N4607V and based the aircraft at Sarasota, Florida. Smith painted the fighter in a bright orange and red paint scheme and made a number of modifications, including a P-51D radiator scoop, doghouse, and radiator. (Doug Fisher Collection)

P-51A-1-NA 43-6006

Built: Inglewood, California; March 1943

Service

1943–44: Delivered to and served at Ladd Field, Alaska.
1944: On May 16, the aircraft crashed near Summit, Alaska, killing Lt. Edward W. Getter.
1977: It was recovered and restoration began by Waldon D. "Moon" Spillers.
1985: On July 3, it had its first post-restoration flight.
1995: It was sold to Jerry Gabe, San Jose, California.
2012: It was acquired by John J. Dowd, Jr.
Current Owner: John J. Dowd, Jr., Syracuse, Kansas (N51Z)

Current Scheme: *Shanty Irish*

The wreck of 43-6006 was recovered in spring 1977 by Waldon D. "Moon" Spillers. He spent more than a decade restoring the fighter, incorporating a set of P-51D wings. Registered N51Z, the aircraft made its first post-restoration flight on July 3, 1985. A decade later, 43-6006 was acquired by Jerry Gabe, who had artist Rick Ruhman paint the nose art of *Polar Bear* to honor the aircraft's service in Alaska. Gabe took the Mustang to the Reno Air Races for a number of years, with Dave Morss handling the piloting duties. Gabe sold the aircraft to Warren Pietsch in 2012, who in turn sold it to John J. Dowd, Jr. New owner Dowd wanted the P-51 gone-through to ensure the fighter was safe to fly. What was planned as an inspection by Pacific Fighters of Idaho Falls, Idaho, turned out to be a major restoration.

Below left: An outstanding study of P-51A 43-6006 in the skies over Idaho. (Jim Raeder)

Below: Cockpit of the restored P-51A is compact with the modern radios hidden to maintain the fighter's World War Two-era appearance. (Jim Raeder)

John Dowd had the fighter re-restored by Pacific Fighters of Idaho Falls, who returned the aircraft in a scheme resembling how the highly polished prototype XP-51 would have looked when flying at Wright Field during the war. (Jim Raeder)

P-51A-10-NA 43-6251

Built: Inglewood, California; April 1943
Service
1943–45: It spent its entire USAAF career as a trainer at various bases in Florida, including the 901 Base Unit (BU) Orlando Army Air Field, to Zephyrhills Army Air Field, to Leesburg Army Air Field, back to Zephyrhills Army Air Field, to 901 BU Orlando Army Air Field, to 904 BU Kissimmee Army Air Field, to 2541 BU Maytag Army Air Field, to 2539 BU (Fighter Gunnery School) Foster Army Air Field.
1945: On May 14, it went into storage at Albuquerque, New Mexico.
1946–55: It moved from War Assets Administration to Grand Central Aircraft Co. (technical school), Glendale, California.
1955 to present: On October 12, it was sold by Grand Central Aircraft Co. to The Air Museum, along with four other aircraft (P-40N, P-47G, P-63A, Me-262). It was registered as N4235Y in 1981.
Current Owner: The Air Museum, Chino, California (N4235Y)

Current Scheme Honors:
Maj. Robert L. Petit's P-51A, *Miss Virginia*.

Left: Mustangs of the 1st Air Commando Group fly over the Chin Hills in Burma while escorting a B-25 strike against Japanese military targets. *Miss Virginia* was flown by Maj. Robert L. Petit. (National Archives)

Below: P-51A *Miss Virginia* taxies out at Chino, California, home to the Planes of Fame Air Museum. Museum founder Ed Maloney recovered this rare fighter from a technical school in Glendale, California, in 1955. (Nicholas A. Veronico)

John Maloney flies P-51A N4235Y over the central California hills, east of San Francisco during the Tracy, California, airshow. The Mustang wore an RAF paint scheme and has subsequently been painted to represent a 1st Air Commando Group aircraft. (Nicholas A. Veronico, photo plane pilot: Taigh Ramey)

P-51C-1-NT 42-103293

Built: Dallas, Texas; February 1944
Service
1944: It departed the United States for the Eighth Air Force (code SOXO) on March 19, 1944. The aircraft was lost in a training accident on May 1, 1944, while assigned to 370th Fighter Squadron, 359th Fighter Group. Capt. Carey H. Brown, Jr. perished when he crashed approximately two miles northeast of Knettishall, England. Capt. Brown had flown 100 hours in combat and was credited with one kill.
2002–17: It was restored by Pacific Fighters of Idaho Falls, Idaho, into wartime dual control TP-51C for Max Chapman. In 2008, it was acquired by the Collings Foundation of Stowe, Massachusetts, and campaigned around the United States in the Wings of Freedom Tour, providing living history flight experiences.
2018–22: Stood down from the Living History tours, 42-103293 was sent to American Aero Services in New Smyrna Beach, Florida, for a complete overhaul. In 2022, the two-seat Mustang is nearing the end of its restoration and will emerge as 354th Fighter Group's *The Stars Look Down*, the aircraft flown by Gen. Elwood "Pete" Quesada, which flew Gen. Dwight D. Eisenhower over the D-Day beaches in June 1944.
Current Owner: Collings Foundation, Stowe, Massachusetts. (N251MX)

New Scheme Honors: P-51B-5-NA 43-6877 *The Stars Look Down*
Built: November 1943 at Inglewood, California, salvaged May 1945 in England
Service: 355th Fighter Squadron, Criqueville, France, July 1944

Right: *Betty Jane* escorts the Collings Foundation's B-24J *Witchcraft* during a training mission at Bomber Camp, hosted by the Stockton Field Aviation Museum in Stockton, California. At the camp, participants are taught bombing and aerial gunnery skills using the actual equipment from World War Two. (Roger Cain)

Left: TP-51 *Betty Jane* sets up for a landing after a Living History flight. (Jim Dunn)

The Collings Foundation owns and operates both TP-51C *Betty Jane* – soon to be flying as *The Stars Look Down* – and a recreation Messerschmitt Me-262, shown in the colors of Luftwaffe I Gruppe of KG51. (David Leininger)

P-51C-5-NT 42-103645

Built: Dallas, Texas; April 1944

Service

1944–45: It first saw service with the Third Air Force, Clearwater, Florida, then went to the 339th Base Unit (Combat Crew Training School – Fighter) at Thomasville Army Air Field, Georgia, and was then surplused on October 3, 1945, to Montana State College, Mechanical Engineering Department.

1945–93: It was at Montana State College before being acquired by F. Robert May. In December 1993, it transferred to the American Airpower Heritage Foundation (Commemorative Air Force – CAF).

2001: In April, it was transferred to Tri-State Aviation, Whapeton, North Dakota, for restoration.

Current Owner: American Airpower Heritage Foundation, Midland, Texas, (N61429), operated by the Commemorative Air Force Red Tail Squadron, Red Wing, Minnesota.

Current Scheme Honors:
All African-American service men and women focusing on the 332nd Fighter Group, known as the Tuskegee Airmen.

Left: P-51C 42-103645 and B-17F 42-3470 shortly after arrival at Gallatin Field, for use by Montana State College's Mechanical Engineering Department. (Galvin Flying Service via Scott Thompson)

Right: Red Tails in Action: The CAF's P-51C in formation with Kermit Weeks' P-51C *Ina the Macon Belle*. (Richard H. VanderMuelen)

The paint scheme on 42-103645 represents an aircraft as flown with the 332nd Fighter Group, known as the Tuskegee Airmen, when operating in the Mediterranean Theater of Operations during World War Two. (David Leininger)

P-51C-10-NT 42-103831

Built: Dallas, Texas; May 1944
Service
1944–45: It served as a trainer at various Third Air Force bases, all of which were in Florida. On October 5, 1945, it was flown to the RFC depot at Stillwater, Oklahoma.
1946: It was sold to a consortium led by Paul Mantz, who acquired the entire storage field at Stillwater on February 19, 1946. It was registered as NX1204 to Paul Mantz Air Services.
1986: On January 25, it was sold to Kermit Weeks at the Tallmantz Auction.
Current Owner: Kermit Weeks, Polk City, Florida (NX1204)

Current Scheme Honors:
Lt. Col. Lee A. "Buddy" Archer, Jr., pilot with the 332nd Fighter Group's 302nd Fighter Squadron. Archer flew 169 combat missions and was credited with four aerial victories – three Hungarian Bf 109s, shot down on October 12, 1944.

Opposite: Kermit Weeks had his P-51C restored in the markings of Lt. Col. Lee A. Archer, Jr. Archer named his Mustang *Ina the Macon Belle* for his wife Ina Burdell. (David Leininger)

Right: Cockpit of *Ina the Macon Belle*. (Richard H. VanderMuelen)

Left: Pilot of the original *Ina*, Lee Archer, autographed the pilot's seat armor plate: "The Macon Belle (Ina) flies again. 01-19-01. Thank you, Kermit." (Richard H. VanderMuelen)

P-51B-10-NA 42-106638

Built: Inglewood, California; February 1944
Service
1944: It departed the United States for the Eighth Air Force on March 16, 1944, code SOXO (Eighth Air Force, England). It served with 376th Fighter Squadron, 361st Fighter Group at Bottisham, England. While being flown on June 22, 1945, by Flight Officer Wade C. Ross, the engine caught fire and Ross parachuted out. The aircraft impacted five miles south of Downham Market at Little Walden, and the aircraft was destroyed.
2002: The aircraft wreckage recovery began.
2005: The remains were sold by Craig Charleston to John T. Sessions Historic Aircraft Foundation, Seattle, Washington. Restoration began at Pacific Fighters.
2008: First post-restoration flight was on July 17. It won Best P-51 and the Silver Wrench Award at EAA AirVenture 2008.
Current Owner: Historic Flight Foundation, Spokane, Washington.

Opposite: Fitted with a Malcolm Hood to provide improved visibility, *Impatient Virgin?* took to the skies 63 years after she went down during a training flight over England.

Right: Cockpit of *Impatient Virgin?* features a Mark II reflector gunsight, typically found in Royal Air Force Spitfires.

Left: Vargas girl pin-up artwork on *Impatient Virgin?*. (All photos by David Leininger)

P-51B-1-NA 43-12252

Built: Inglewood, California; September 1943

Service

1944–45: It was delivered to the USAAF on September 29, 1943, and assigned to the Third Air Force, Bartow Army Air Field, Florida. On November 14, 1944, it crashed into Lake Louisa (west of Orlando), Florida. Pilot 1st Lt. Dean R. Gilmore perished in the crash.

2002–08: The wreckage was purchased by Jack Roush in 2002, registered as N551E, and shipped to Cal Pacific Airmotive in Salinas, California, for restoration. It was restored in the colors of Col. C. E. "Bud" Anderson's *Old Crow*.

2008: On June 8, the first post-restoration flight was with Dan Martin at the controls.

Current Owner: Jack Roush, Livona, Michigan (N551E)

Current Scheme Honors:

P-51B 43-24823 flown by triple ace Col. C. E. "Bud" Anderson.

Built: March 1944, service with the Eighth Air Force, 357th Fighter Group at Leiston Airfield, England.

Opposite: The late Jimmy Leeward at the controls of *Old Crow* over Lake Winnebago flying during EAA AirVenture. (David Leininger)

Right: Note how the Malcolm Hood improves visibility for the pilot, as well as the rearview mirror above the cockpit. (David Leininger)

Left: During Anderson's second tour, he exclusively flew this P-51D 44-14450, downing four additional German aircraft from this mount. (Courtesy Col. C. E. "Bud" Anderson)

P-51C-10-NT 43-25057

Built: Dallas, Texas; June 1944

Service

1944–46: It was delivered to 568th BU (Operational Training Unit), Brownsville, Texas; the unit then moved to Greenwood, Mississippi, and then to 4160 BU, Hobbs Air Field, New Mexico. On April 6, 1946, it was transferred to RFC Ontario, California.

1946: Later in the year, it was purchased from the RFC by Jack Hardwick, El Monte, California.

1979 to present: John R. Paul purchased the hulk from Jack Hardwick on March 16, 1979; March 26, 2009, initial registration, N4651C, John R. Paul, Meridian, Idaho; first post-restoration flight was in 2010. On June 25, 2012, ownership changed to Mustang LLC, Meridian, Idaho.

Current Owner: Mustang LLC, Meridian, Idaho

Based: Warhawk Air Museum, Nampa, Idaho

Current Scheme Honors:
Lt. Col. Duane W. Beeson, 334th Fighter Squadron, 4th Fighter Group, England.

Opposite: *Boise Bee* was the personal mount of Duane W. Beeson, who was born and raised in Boise, Idaho. The Mustang was painted with 20 of Beeson's eventual score of 22.08 aerial victories. (Jim Raeder)

Right: Photographer William T. Larkins captured 43-25057 sitting at RFC Ontario, which is today's Chino Airport. Very few aircraft at RFC Ontario escaped the scrapper's torch, so a photo of this aircraft is extremely rare. (William T. Larkins)

Left: Lt. Col. Duane W. Beeson achieved 17.33 aerial victories while flying the P-47 Thunderbolt and the balance in the P-51 Mustang. He was born and raised in Boise, Idaho, near the home of the Warhawk Air Museum. (National Archives)

P-51C-10-NT 43-25147

Built: Dallas, Texas; June 1944

Service

1944–45: On October 9, 1944, it departed the United States for the 10th Air Force, Calcutta, India (code DAUB); On August 7, 1945, it joined the 20th Tactical Reconnaissance Squadron in India.

1970s–1981: Aircraft restorer Peter Regina traveled to Israel to recover a number of P-51s. He found a P-51B wing and, reconstructing the fuselage of Israeli Air Force P-51D IDF-13 as a B model along with the tail section of P-51B 43-6351 (N68738), was able to build a P-51B, registered in March 1981 as N51PR, which flew in the colors of *Shangri-La*.

1986: It was sold to Joseph Kasparoff, Montebello, California. The Mustang was flown as *The Healer* and *Super K*.

1998: It was sold to Stephen Grey's Patina Ltd., Jersey, Channel Islands, United Kingdom, with a deregistration request from the US Civil Register; UK registration G-PSIC. The aircraft's paint scheme was subsequently changed to that of *Princess Elizabeth*.

2006: It was purchased by Jim Beasley's HMBP51C, LLC, Coatsville, Pennsylvania, and registered N487FS.

2008: On June 13, 2008, it was acquired by Dan Friedkin's Comanche Warbirds, LLC, Houston, Texas. *Princess Elizabeth*.

Current Owner: Comanche Warbirds, Inc. Houston, Texas (N487FS)

Current Scheme Honors:
Capt. William T. "Bill" Whisner, 487th Fighter Squadron, 352nd Fighter Group's P-51B 42-106449, *Princess Elizabeth*, which was lost on D-Day, June 6, 1944, with Lt. Robert K. Butler at the controls. Butler had been strafing a German convoy five miles west of Pont-Audemer, France, when his aircraft was hit by ground fire in the engine and fuselage tank. He baled out and successfully evaded capture, returning to Allied lines.

Above: Comanche Warbirds' A-36 and P-51C *Princess Elizabeth* shows the Allison engine installation as compared to the Merlin engine in the C model. Notice the absence of the fuselage fillet ahead of the vertical stabilizer, as seen on later model Mustangs. (All photos by Richard VanderMuelen)

Right: Cockpit of *Princess Elizabeth*.

The original *Princess Elizabeth* wore invasion stripes, as seen in this paint scheme, on the day the aircraft was lost over France. This view shows the C model's wing guns, two per side, to advantage.

P-51D-5-NA 44-13521

Built: Inglewood, California; May 1944
Service
1944: This Mustang arrived in England on June 6, 1944, and was assigned to the 504th Fighter Squadron, 339th Fighter Group, Eighth Air Force at Fowlmere, Cambridgeshire. The P-51D was assigned to Capt. Bradford V. Stevens, who named the Mustang *Marinell* after his former girlfriend Mary Anell (Stokes) Shuffleton. Stevens scored two of his four aerial victories in *Marinell*. During the afternoon of August 13, 1944, Lt. Myer Winkleman flew *Marinell* on a dive bombing and strafing run to the Beauvais area in France. Winkleman perished when the Mustang crash near Feuquieres, France.
1945–88: The aircraft wreckage was stored, and eventually displayed in a museum. Subsequently, the Mustang wreckage was purchased by Maurice Hammond.

2003: Hammond began restoration, and it was first flown on July 26, 2008, as G-MRLL.
2017: On May 17, it was sold to Sky West Aviation Inc., Albuquerque, New Mexico (N383FJ), on behalf of Carlo Coltri, Collezione Marchi, Luciano Sorlini, Italy.
Current Owner: Sky West Aviation Inc., Albuquerque, New Mexico (N383FJ)

Opposite: Combat veteran P-51D 44-13521 lifts off from the grass runway. Capt. Stevens' markings show four aerial victories. (Mike Shreeve)

Right: *Marinell* was named for Capt. Stevens' former girlfriend. (Mike Shreeve)

Left: Capt. Bradford V. Stevens of the 339th Fighter Group is presented a medal by Brig. Gen. Murray C. Woodbury at Eighth Air Force Station F-378 in England on July 22, 1944. (71764AC, National Archives)

P-51D-20-NA 44-63577

Built: Inglewood, California; November 1944

Service

1944–49: It served stateside at Birmingham and Mobile, Alabama, before moving to storage at Kelly AFB, Texas.

1949–85: On April 7, 1949, it transferred to the Uruguayan Air Force, and received serial number FAU-265. In 1960, it was displayed at the Museo de Aeronautica, Montevideo, Uruguay. It was acquired by Danta Heredia, of Montevideo. It was then sold to Tyrone Elias of Tulsa, Oklahoma, and was registered as N51TE.

1994–2012: On July 15, 1994, the aircraft was acquired by John R. Turgyan of New Egypt, New Jersey. The restoration project moved to Midwest Aero Restorations, Danville, Illinois, for completion.

2013: First post-restoration flight was on July 11, 2013. Jon S. Vesely, Scottsdale, Arizona, acquired the project on September 13.

2014 to present: On August 7, 2014, it was sold to Lawrence Classics LLC, Bentonville, Arkansas. On April 9, 2020, ownership changed to Echo Matrix LLC, also in Bentonville.

Current Owner: Echo Matrix LLC, Bentonville, Arkansas (N151JT)

Current Scheme Honors:
Aircraft flown by 12th Fighter Squadron, 18th Fighter Bomber Wing, Chinhae Airfield, South Korea.

Opposite: All of the original armament parts, including the stub rocket rails, were acquired from now-retired warbird parts dealer Jay Wisler. The rocket rails were manufactured on metal sheets, and once the correct rivets in the underside of the wing were removed, the rails screwed in perfectly.

Right: Details of the cockpit interior, port side, showing the level of detail provided by Midwest Aero Restorations of Danville, Illinois.

Left: Port side marking of *Was That Too Fast* showing the 12th Fighter Squadron emblem, a shark's mouth, and aircraft data block. (All photos by David Leininger)

P-51D-20-NA 44-63675

Built: Inglewood, California; November 1944
Service
1944–45: The aircraft departed the United States in February 1945. It was assigned to the 402nd Fighter Squadron, 370th Fighter Group of the Ninth Air Force, based at Ophoven Airfield, Belgium. It was flown by Lt. Robert "Bob" Bohna. After the war, it was stored at Fürth, Germany.
1947: It went to the Swedish Air Force as serial Fv26152, named *Gul Kalle* (*Yellow K*).
1954–70: In September 1954, it went to the Nicaraguan Air Force (Fuerza Aérea de la Guardia Nacional), reported as assigned serial number GN-91. Retired in 1961, it was displayed at the Officers Club, Managua Air Base.
1970–2011: It was recovered by Dave Allender and was the second aircraft to wear registration N5452V. First flight in the United States was on September 11, 1973. It was raced at Reno as number 19. It was sold to Roger Christgau in 1977 (N1751D), and then sold to Paul Ehlen in July 2011.
2011 to present: Ehlen contracted Air Corps Aviation, Bemidji, Minnesota, for restoration work. First post-restoration flight was on September 17, 2014.
Current Owner: Paul Ehlen, Rare Birds LLC, Bloomington, Minnesota (N1751D)

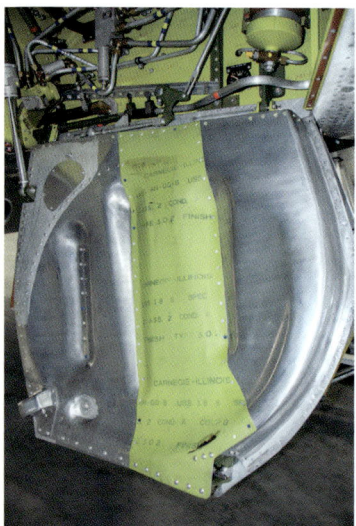

Opposite: P-51D-20-NA 44-63675 was flown in combat over Germany by Lt. Robert "Bob" Bohna. After World War Two, 44-63675 flew with the Swedish Air Force, then with the Nicaraguan Air Force, before being returned to the United States in 1970. The aircraft was restored by Air Corps Aviation in Bemidji, Minnesota. (Richard H. VanderMeulen)

Right: Kevin "Max" Wisniewski adds the finishing touches to *Sierra Sue II*'s artwork on the Air Corps Aviation ramp. (Richard H. VanderMeulen)

Left: Clamshell door and gear well interior showing the level of detail undertaken in this restoration. (David Leininger)

P-51D-20-NA 44-63807

Built: Inglewood, California; December 1944

Service

1944–45: US Eighth Air Force Fighter Command (squadron unknown).
1945–49: It served with various USAAF and USAF Base Units from October 1945, until it was dropped from inventory in November 1949.
December 1950–March 1960: The Mustang saw service with the Uruguayan Air Force as FAU 272.
1960–77: The aircraft served with the Bolivian Air Force (Fuerza Aérea Boliviano, FAB) as FAB 506.
1977: It was exported to Canada and became C-GXUO.
1985–2004: It passed through various owners in the United States.
Current Owner: PACWEST P-51 LLC, Sacramento, California (N20MS)

Current Scheme Honors:

P-51D-10-NA 44-14733 *Daddy's Girl*
Built: Inglewood, California; 1944
Service: US Eighth Air Force Fighter Command 370th Fighter Squadron of the 359th Fighter Group.
Pilot: Capt. Raymond S. Wetmore, who had 21.25 kills, nine of which were in 44-14733 *Daddy's Girl*.

Opposite: Twenty-three aerial victory marks adorn the port side of *Daddy's Girl*. (Jim Dunn, photo plane pilot: Jerry Anderson)

Right: Cockpit detail of *Daddy's Girl*. (Jim Dunn)

Left: Known as "The Man with Telescopic Vision," Capt. Ray S. Wetmore, age 22, from Kerman, California, talks with his armorer, Sgt. Locklyn Sangster, right, after shooting down three Nazi aircraft. (National Archives)

P-51D-20-NA 44-63864

Built: Inglewood, California; December 1944

Service

1945–48: Departed the United States on February 28, 1945, for the Eighth Air Force in England. It was assigned to the 83rd Fighter Squadron, 78th Fighter Group stationed at Duxford, England. It was flown by 1st Lt. Hubert "Bill" Davis, who named the Mustang *Twilight Tear*. Stored in England following the conclusion of hostilities, the Mustang was sold to the Swedish Air Force in 1948 and assigned Swedish Air Force serial J26.

1953–60: It was sold to the Israeli Defense Force Air Force (IDFAF) and assigned serial number 2338. In 1958, it was removed from service.

1960–2005: It was sold to William P. Lear and modified with Cavalier tip tanks. On June 6, 1963, it crashed at Keflavik, Iceland, on a ferry flight from Geneva to the United States. It was then dumped at Keflavik. Next, it was acquired by Cham Gill of Oregon in 1989 (registered N42805), and the wreck was moved to United States. The Mustang project was subsequently acquired by Kenneth Hake.

2005: On October 14, it was sold to Ron Fagen, Granite Falls, Minnesota. Restoration began the same year.

Current Owner: Fagen Fighters WWII Museum, Granite Falls, Minnesota (N251L)

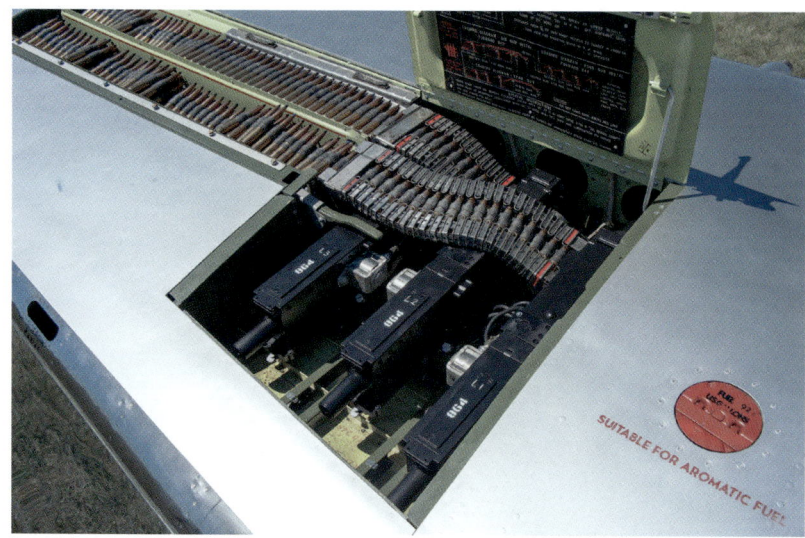

Opposite: Twilight Tear was flown in combat by 1st Lt. Hubert "Bill" Davis, who was credited with four aerial victories during the war.

Right: Port wing machine guns loaded and ready. Notice the gun firing solenoids, as well as the ammunition loading chart on the inside of the gun bay door.

Left: Port side details of *Twilight Tear*. This aircraft has had fully functional 0.50-cal. machine guns installed for a period of time. (All photos by David Leininger)

P-51D-20-NA 44-72035

Built: Inglewood, California; December 1944

Service

1944–45: 385th Fighter Squadron, 364th Fighter Group (fuselage code 5E * B), Eighth Air Force; subsequently went to 332nd Fighter Group, 15th Air Force; it returned to United States and was stored.
1947–53: ANG service in Pennsylvania and, later, Oklahoma.
1956: Entered storage at McClellan AFB, Sacramento, California.
Sept. 25, 1957: Sold in McClellan sale to Whiteman Enterprises, Pacoima, California, for US$1,100. The aircraft had 1,083.15 hours total time on the airframe. It was registered as N5411V.
1981: Sold to H. Escobar, Bogotá, Columbia, and subsequently registered as HK-2812P.
1988: Sold to Jacques Bournet in France and registered as F-AZMU, *Jumpin' Jacques*.
2002: Sold to Peter Teichman, and subsequently registered as G-SIJJ.
2016: Repainted in its wartime colors of *Tall in the Saddle*.
Current Owner: Peter Tiechman, Hangar 11 Collection, North Weald, Essex, England (G-SIJJ)

Below: View of the 99th Fighter Squadron's Red Tail markings on *Tall in the Saddle*, which was flown by Capt. Wendell Lucas and Lt. Col. George Hardy. (Photo by Mike Shreeve)

Current Scheme: *Tall in the Saddle*

On January 4, 1945, Inglewood-built 44-72035 was delivered to the Port of Embarkation at Newark, New Jersey, and shipped to the European Theater of Operations. The fighter was first assigned to the 385th Fighter Squadron, 364th Fighter Group, at Honington, Suffolk, England. It was reportedly assigned to the Tuskegee Airmen of the 332nd Fighter Group.

Left: Tuskegee Airman Edward C. Gleed in front of his P-51D *Creamer's Dream* at Ramitelli, Italy, in March 1945. Gleed flew with the 99th Fighter Squadron and was the operations officer for the 332nd Fighter Group. He downed a pair of Fw 190s over Budapest, Hungary, on July 27, 1944. *Tall in the Saddle* pays tribute to Gleed and the other members of the 332nd Fighter Group. (Library of Congress)

Tall in the Saddle takes off. Notice the clamshell gear doors closing over the wheels. (Mike Shreeve)

P-51D-20-NA 44-72051

Built: Inglewood, California; January 1945

Service

1945–51: Delivered to USAAF in January 1945. It went to the Ninth Air Force, and then to the Swedish Air Force as Fv26026.

1952–84: It was sold to the Dominican Air Force (Fuerza Aérea de República Dominicana) as FAD 1912 in October 1952.

1984–95: On May 19, 1984, it was recovered by Brian O'Farrell. It was then sold to John R. Sandberg in 1985, and restored as *Platinum Plus*. It raced at Reno as number 28. In 1991, it was sold to Janet S. Bjornstad.

1995 to present: On July 26, 1995, it was sold to Ron Fagen, and restored as *Sweet Revenge*.

Current Owner: Fagen Fighters WWII Museum, Granite Falls, Minnesota

Current Scheme Honors:

Aircraft of the 334th Fighter Squadron, 4th Fighter Group. The 334th was the successor to No. 71 Eagle Squadron, RAF, which was comprised of American pilots.

Right: *Sweet Revenge* on the ramp at Fagen Fighters WWII Museum in Granite Falls, Minnesota. (Richard H. VanderMeulen)

Left: The red nose with red and white checkered band represents the 334th Fighter Squadron. (Jim Dunn)

Sweet Revenge wears the boxing eagle of the 4th Fighter Group on the port side fuselage under the aircraft's kill tally. (Richard H. VanderMeulen)

P-51D-25-NA 44-72777

Built: Inglewood, California; February 1945
Service
1944–45: Served with the 5th Fighter Squadron, 52nd Fighter Group, 15th Air Force and was the personal mount of five victory ace Major Ralph J. "Doc" Watson.
1946–59: Served with various ANG units, including Rhode Island and California.
1959–79: It was sold surplus to Trans Florida Aviation, Sarasota, Florida. In 1967, it went to the Indonesian Air Force as F-344. In 1971, under Operation *Peace Pony I*, it was modernized with a Cavalier mod kit for the Indonesians.
1979–84: The Indonesian Mustang fleet was recovered by Stephen Johnson, Ralph S. Johnson and Art Stagg of Oakland, California, who sold them as restoration kits. It was acquired by Al Letcher and registered as N8064V. It was assembled and flown on April 22, 1981, and named *Singapore Sally*.
1984 to present: Acquired by Steve Seghetti, Vacaville, California, it was reregistered as N151D. It was subsequently painted as *Sparky*. It raced at Reno as number 44, *Jelly Belly*, and most recently turned the pylons wearing the markings of *Blondie*.
Current Owner: Brant Seghetti, Vacaville, California

Current Scheme Honors:
P-51D-25-NA 44-73304 *Blondie*, 334th Fighter Squadron, 4th Fighter Group, at Debden, England, flown by 2nd Lt. Marvin W. Arthur. It was named for Arthur's wife.

Opposite: Wearing *Sparky* nose art, Brant Seghetti flies 44-72777 over the farm fields near Vacaville, California. (Jim Dunn)

Right: Artwork on 44-72777 *Blondie* is sexier than the wartime original. (Jim Dunn)

Left: After service with the Rhode Island ANG, 44-72777 was transferred to the California ANG and is seen at Hayward, California. This aircraft was sold in the McClellan auctions and had accumulated 1,892.30 hours total time on the airframe during its military service. It did not sell in the first auction. (William T. Larkins)

P-51D-25-NA 44-73264

Built: Inglewood, California; March 1945

Service

1945–57: Sent overseas in March 1945, it returned stateside in July 1945. It served with various Army Air Forces units in New Jersey and Pennsylvania, before being transferred to the Wyoming ANG in March 1947. Subsequently, it served with the Illinois, Kentucky, and New Mexico ANGs. In 1956, it was sent to McClellan AFB for storage.

1957: It sold at the McClellan auction for US$700.70 to Abraham Miller of Baltimore, Maryland. The aircraft had accumulated 1,574.55 hours total time airframe and 527.50 hours on the engine. It was registered as N5428V.

1957–70: In succession, the Mustang was owned by Mathew Kibler, Charles Schalebaum, John Sliker, and William Ross, before passing to the Confederate Air Force.

1970 to present: It is owned and flown by the Commemorative Air Force. The long-time sponsor of this aircraft was Brig. Gen. Regis F. A. Urschler (USAF ret.).

Current Scheme Honors:
Aircraft of the 343rd Fighter Squadron, 55th Fighter Group, Eighth Air Force, based at Nuthampstead (September 1943–April 1944) and Wormingford (April 1944–July 1945), England.

Opposite: *Gunfighter* took to the skies during a recent EAA AirVenture fly-in convention. The aircraft was heavily damaged in a 1981 crash, and, when rebuilt, emerged in this scheme honoring the men and aircraft of the 343rd Fighter Squadron. (David Leininger)

Right: Reg Urschler in a pre-1981 paint scheme *Gunfighter*. (Jerry Liang)

Left: *Gunfighter* nose art. (David Leininger)

P-51D-25-NA 44-73343

Built: Inglewood, California; March 1945

Service

1945: Delivered to Hobbs Army Air Field, New Mexico, followed by storage at Kelly Army Air Field, Texas (dates and other details unreadable on card).

1957–98: Purchased at the McClellan auction for US$1,400 by Ben Hall of Seattle, Washington. It was registered as N5482V and was flown as both *Seattle Miss* and *Esther's Mink*. In 1967, it was sold to Michael Loening. It crashed at the 1971 Reno Air Races. Stored until 1978, it was then acquired by Bruce Morehouse.

1998–2017: The Mustang was delivered to Midwest Aero Restorations, Danville, Illinois, for restoration. The project was then sold to Larry Thompson in 2005, and onto Jon Vesley in 2008. First post-restoration flight was in August 2010. It was registered N551JV. Restoration awarded both the Reserve Grand Champion WWII Warbird and Golden Wrench Award (for Midwest Aero Restorations) at EAA AirVenture 2010.

2017–20: It was sold to Rod Lewis, Lewis Fighter Fleet, LLC, San Antonio, Texas.

2020: Acquired by Charles Somers, Hillsboro, Oregon.

Current Owner: Charles Somers, Hillsboro, Oregon, based at McClellan Field, Sacramento, California

Current Scheme Honors:
Capt. Clayton K. Gross' 44-63668 *Live Bait* of the 355th Fighter Squadron, 355th Fighter Group.

Opposite: *Live Bait* on one of its early outings in the skies over Wisconsin.

Right: Port side fuselage markings on *Live Bait*. Notice the partial kill flag. Capt. Clayton K. Gross was credited with 6.5 aerial victories (six Bf 109s confirmed, three Bf 109s damaged, and an unconfirmed kill of an Me-262 in the area around Lonnewitz, Germany, on April 14, 1945).

Left: Radiator piping details under the skin of *Live Bait*.
(All photos by David Leininger)

P-51D-25-NA 44-73656/44-12473

Built: Inglewood, California; October 1944

Service

1945–47: Shipped to the Eighth Air Force in England in October 1944, it was returned to the United States in July 1945 and stored. In March 1947, it went to the 182nd Fighter Squadron, Texas ANG.

1950–58: In November 1950, it moved to the 136th Fighter Group, Langley AFB, Virginia; in May 1951, to the 120th Fighter Squadron, Colorado ANG; in March 1953, to the 109th Fighter Squadron, Minnesota ANG; and in December 1956, to McClellan AFB for storage.

1958–68: It was sold surplus to Delta A&E Parts for US$1,307.50. In December 1958, it went to Trans Florida Aviation, and was converted into a Cavalier 750 executive transport. Then, in 1963, it was sold to Howard Olsen, Midland, Texas. In April 1968, it moved onto Duncan Airmotive, Galveston, Texas. It was then sold to the El Salvador Air Force (Fuerza Aérea Salvadoreña).

1974–88: Returned to the United States by Jack Flaherty, its serial number was changed to 44-12473. The Mustang was then acquired by Gordon Plaskett of King City, California, in 1975, and restored as *Moonbeam McSwine*. It was then acquired by Chris Williams of Ellensburg, Washington, in 1987.

1988–2017: It was sold to Vlado Lenoch in April 1988. Exported to France, it went to Frederic Akary and was registered F-AZXS.

2018 to present: Sold to Paul Wood/Moonbeam Historic Military Aircraft and returned to the United States.

Current owner: Paul Wood/Moonbeam Historic Military Aircraft LLC., Waukegan, Illinois (N51VL)

Current Scheme Honors:

Triple ace (15.5 aerial victories) Capt. William T. Whisner's *Moonbeam McSwine* of the 487th Fighter Squadron, 352nd Fighter Group, based at Bodney, England.

Right: Vlado Lenoch taxies in at EAA AirVenture after an outstanding aerial demonstration of the Mustang's capabilities. (Jim Dunn)

Left: *Moonbeam McSwine* and *Excalibur* (45-11540, N151W) both in the colors of the 352nd Fighter Group, known as the "Blue-nosed Bastards of Bodney," in the Mustang Corral at EAA AirVenture. (Jim Dunn)

Paul Wood's *Moonbeam McSwine* wears the registration N51VL in tribute to the late Mustang pilot and aviation educator Vlado Lenoch. (David Leininger)

P-51D-25-NA 44-74202

Built: Inglewood, California; May 1945

Service

1945–48: Served at various stateside bases, including Bakersfield, Santa Maria, and March Fields in California. It was sent to storage at Hobbs Army Air Field, New Mexico, in February 1946. A storage transfer to Kelly AFB, Texas, followed in June 1947.

1948–57: It was removed from storage and assigned to the 27th Fighter Group (Strategic Air Command), Kearney, Nebraska, followed by assignments with the Colorado ANG, New Mexico ANG, and Kentucky ANG. It was then sent to McClellan AFB for storage in October 1956 and was sold surplus in the 1957 McClellan auction.

1958–2007: It was sold surplus to Earl Dakin of Sacramento, California, for US$920, and registered N5420V. Next it went to Mike Coutches in 1966. It was stored until 1984, when it was sold to Mike Bogue of Oakland, California. Mike Coutches then reacquired it in 1990. It was registered to Robert Coutches in 1999.

2007: It was sold to Jack Croul. First post-restoration flight was in May 2012.

2012 to present: It was sold to Robert Dickson, Jr. In 2013, it was awarded Best Fighter at EAA AirVenture with Vintage Airframes, Caldwell, Idaho, taking home the Silver Wrench Award.

Current Owner: Robert Dickson, Jr., Fox51 LLC, Concord, North Carolina

Current Scheme Honors:
Lt. Col. William W. "Will" Foard's P-51D 44-15660 *Swamp Fox*, 357th Fighter Group, based at RAF Leiston, England.

Opposite: Wearing the colors of the 357th Fighter Group, *Swamp Fox* shines in the afternoon Minnesota sky. This Mustang was restored by Vintage Airframes of Caldwell, Idaho. (David Leininger)

Right: *Swamp Fox* with drop tanks on the wing hard points. (Roger Cain)

Left: In the Mustang Corral at EAA AirVenture, the polish work on *Swamp Fox* shines brightly. (Roger Cain)

P-51D-30-NA 44-74391

Built: Inglewood, California; May 1945

Service

1945–47: Accepted by the USAAF on May 23, 1945, it was immediately sent to storage.

1948–58: It went to the 113th Fighter Squadron, Indiana ANG; in 1950, it moved to the North Carolina ANG; in November 1950, it went to the 123rd Fighter Bomber Wing, Tactical Air Command, Goodman AFB, Kentucky; December 1951 saw it go to the 146th Fighter Bomber Wing, George AFB, California; in January 1953, it then went to the 21st Fighter Bomber Squadron, George AFB; in May 1953, it transferred to the California ANG; June 1955, it moved to the Illinois ANG; and in October 1956, it was sent to storage at McClellan AFB, California.

1958–72: It was acquired by the Guatemalan Air Force (Fuerza Aérea Guatemalteca, FAG) as serial number FAG 351.

1972–2009: It was purchased in August by Don Hull, Sugarland, Texas, and registered as N38229. In November 1973, it was acquired by Wilson C. "Connie" Edwards, Big Spring, Texas. Then, it was purchased by Herber Costello in June 2000. It was sold to Woods Aviation, Carefree, Arizona, in October 2000, and registered as N351MX. It was restored as ace Jim Brooks' *February*. Subsequently, it was painted as *The Hun Hunter~Texas*.

2009: It was acquired by Comanche Fighters of Houston, Texas. In 2018, it was shipped to Duxford, England, for the airshow season.

Current Owner: Comanche Fighters LLC., Houston, Texas (N351MX)

Current Scheme Honors:

Capt. Henry W. Brown's P-51D-5-NA 44-13305 *The Hun Hunter~Texas*, in which he was shot down by flak near Nördlingen, Germany, on October 3, 1944. Brown's wingman Chuck Lenfest landed nearby to pick up Brown, but his Mustang became stuck in mud and both pilots were taken prisoner. Brown was credited with 14.2 aerial victories.

Left: Capt. Henry W. "Baby" Brown, left, and Lt. Col. Claiborn H. Kinnard, Jr. of the 355th Fighter Group, with Brown's P-51B-10-NA Mustang (serial number 42-106448) also named *The Hun Hunter~Texas*. (National Archives)

Right: Close-up of the markings on *The Hun Hunter~Texas*. (Mike Shreeve)

The gear comes up as *The Hun Hunter~Texas* lifts off from an airfield in England in July 2021. (Mike Shreeve)

P-51D-30-NA 44-74427

Built: Inglewood, California; June 1945
Service
1945–50: June 1950, delivered to Mabry Army Air Field, Florida, followed by assignment to Hunter Army Air Field, Georgia, and then to Hobbs Army Air Field, New Mexico. To Kelly Field, Texas, for storage in July 1947. In August 1950, the aircraft was removed from storage and flown to Burbank, California, for overhaul and preparation for transfer to the RCAF.
1950–59: In November, it was delivered to the RCAF, serial number RCAF 9592.
1959–62: it was sold to Trans Florida Aviation, Sarasota, Florida, along with seven other RCAF Mustangs. It received a Cavalier Executive conversion, and was registered as N2251D.
1962–71: It was purchased by North American Aviation, which later became North American Rockwell Corp. It became the personal aircraft of R. A. "Bob" Hoover. On August 9, 1970, it was damaged when the oxygen bottles were over-pressured and exploded.
1971-2011: It was rebuilt as *Miss Coronado*, and later *Nooky Booky IV*. On May 17, 2000, it was sold to JCB Aviation of Nimes, France, and registered F-AZSB. It was then sold to Christian Amara in 2011.
2021: It was sold to The Flying Bulls, Salzburg, Austria.
Current Owner: The Flying Bulls, Salzburg, Austria (F-AZSB)

Current Scheme Honors:
Major Leonard K. "Kit" Carson's P-51K-5-NT 44-11622 *Nooky Booky IV*. Carson had four of his 18.5 victories in this aircraft and was credited with damaging three Me 262s in it as well.

Right: The gear comes up as *Nooky Booky IV* climbs out. Maj. Kit Carson scored the majority of his aerial victories in a P-51D and P-51K. (Mike Shreeve)

Left: Major, later Colonel, Leonard K. "Kit" Carson. (National Archives)

Nooky Booky IV about to touch down at an airshow in England. The fighter is now owned by The Flying Bulls, based at Salzburg, Austria. (Mike Shreeve)

P-51D-30-NA 44-74452

Built: Inglewood, California; May 1945

Service

1945: Delivered June 5, 1945, to Pinellas Army Air Field, Florida, and transferred to Hunter Army Air Field, Georgia, on November 12, 1945. Subsequently transferred to Hobbs Army Air Field, New Mexico, where it served until August 6, 1947, when flown to storage at Kelly Army Air Field, Texas. On August 14, 1950, the aircraft was removed from storage and flown to Burbank, California, for overhaul and preparation for transfer to the Royal Canadian Air Force.

1950–58: Acquired on November 15 by the RCAF, and assigned serial number RCAF 9225. It was struck off charge on April 29, 1958.

1958–62: It was sent to Intercontinental Airways, Canastota, New York.

1962–2001: In March 1962, it was sent to the Guatemalan Air Force, with serial FAG 366. In August 1972, all remaining FAG Mustangs and spare parts were recovered by Don Hull, Houston, Texas (N74109). It was then acquired by Wilson C. "Connie" Edwards of Big Spring, Texas, in 1973. It was stored until August 2001, when acquired by Bruce Winter, San Antonio, Texas.

2003–08: It was sent to Midwest Aero Restorations, Danville, Illinois, for restoration. The finished aircraft includes duplicate paper drop tanks. The aircraft won Grand Champion World War II at EAA Airventure 2008 and earned Midwest Aero Restorations its seventh Golden Wrench Award.

Current Owner: Bruce Winter, FTR ESC LLC., San Antonio, Texas (N74190)

Current Scheme Honors:
Maj. Jack M. Ilfrey's P-51D-5-NA 44-13761 *Happy Jack's Go Buggy*. Ilfrey was credited with 7.5 aerial victories.

Right: Details of the markings on *Happy Jack's Go Buggy*. (David Leininger)

Left: Maj. Jack M. Ilfrey had an amazing flying career during World War Two. An ace in the Mediterranean Theater flying P-38s, he was shot down and evaded capture. After a stint as an instructor in the United States, he returned to England in March 1944, to command the 79th Fighter Squadron. He is seen here with his ground crew. Ilfrey was shot down on June 12, 1944, and again evaded capture. Ilfrey was credited with 7.5 aerial victories. (National Archives)

Former F/A-18 pilot Bruce "Doc" Winters in *Happy Jack's Go Buggy* over Pyramid Lake, Nevada, after being awarded the Rolls Royce Aviation Heritage Trophy in September 2008. Note the paper drop tanks that were recreated for this restoration. (Richard H. VanderMeulen)

P-51D-30-NA 44-74483

Built: Inglewood, California; June 1945

Service

1945–50: Delivered June 9, 1945, to Venice Army Air Field, Florida, and transferred to Hunter Army Air Field, Georgia, on October 30, 1945. Subsequently transferred on November 30, 1945, to Hobbs Army Air Field, New Mexico, where she served until August 6, 1947, when flown to storage at Kelly Army Air Field, Texas. On August 17, 1950, the aircraft was removed from storage and flown to Burbank, California, for overhaul and preparation for transfer to the RCAF.

1950–58: The Mustang was transferred to the RCAF as RCAF 9228.

1958–77: It was sold to James Defuria on July 21, 1958, and was registered as N6523D. It passed through various owners until December 5, 1966, when acquired by George Perez.

1966 to present: Owned by George Perez, it was reregistered as N51GP in 1977.

Current Owner: George Perez, Hangar 121 LLC., Sonoma, California. The aircraft is based at Schellville Airport, California (N51GP)

Current Scheme Honors:
P-51 factory delivery paint scheme.

Opposite: George Perez's Mustang is escorted by Chris Prevost's P-40. Both aircraft are on loan to the North Bay Air Museum at Schellville Airport, California. (Jim Dunn)

Right: Visitors to California's Napa Valley can often see N51GP at the Schellville Airport, parked on The Vintage Aircraft Company's ramp. (Jim Dunn)

Left: Factory-fresh Mustangs being transported across the Pacific Ocean for delivery to frontline units. George Perez's Mustang wears this same scheme. (National Archives)

P-51D-30-NA 44-75009

Built: Inglewood, California; July 1945

Service

1945–57: Flown from the factory directly to storage with the 4126th Base Unit at San Bernardino Army Air Field, California. On October 23, 1947, assigned to Kelly AFB, Texas. On November 19, 1947, assigned to the Strategic Air Command at Kearney AFB, Nebraska, and later Grenier AFB, New Hampshire. On June 12, 1950, 44-75009 was transferred to the North Dakota ANG, and on April 16, 1951, to the Iowa ANG. This was followed by a short stint with the Tactical Air Command at George AFB, California. Subsequent assignments to the Minnesota and Ohio ANG followed. By December 1956, 44-75009 was sitting in storage at McClellan AFB awaiting its fate.

1957–85: It was sold surplus at McClellan AFB to Homer Roundtree, Anderson, California, for US$1,177. Then, it spent time in storage. In 1965, it was registered as N5474.

1985: It was sold to Ted Contri. It emerged from restoration in 1985, registered as N51TC, and named *Rosalie*.

2018: On January 29, it was sold to Rickards Aviation Group LLC., Millsboro, Delaware.

Current Owner: Rickards Aviation Group LLC., Millsboro, Delaware (N51TC)

Current Scheme Honors:
Aircraft and personnel of the Nevada ANG.

Left: Nose art on *Rosalie*. (Roger Cain)

Opposite: When 44-75009 was restored by Ted Contri, the Mustang was finished in a Nevada ANG paint scheme. The Nevada ANG flew the Mustang from 1948 to 1955. (David Leininger)

Right: *Rosalie* is seen at the 2019 EAA AirVenture convention at Oshkosh, Wisconsin. (Roger Cain)

P-51D-25-NT 44-84655

Built: Dallas, Texas; May 1945

Service

1945: Delivered May 27, 1945, to the 341st BU (Gunnery Station) Pinellas Army Air Field, St. Petersburg, Florida. June 5, 1945, transferred to the 2539th (Fighter Gunnery School) Foster Field, Victoria, Texas; with 30-day assignments to Pinellas, 4204 BU at Warner Robins, Georgia, and Hunter Army Air Field, Georgia. On December 3, 1945, flown to Hobbs Army Air Field, New Mexico, for storage.

1950: On September 5, 1950, this Mustang was sent to the Texas Engineering and Manufacturing Co. (TEMCO), to be converted with a full dual-control rear cockpit.

1952: It served with the 120th Fighter Bomber Squadron, 140th Fighter-Bomber Group, Colorado ANG.

1954: It was transferred to the 167th Fighter Squadron, West Virginia ANG.

1956: It was flown to McClellan AFB, California, for storage and disposition.

1958: On February 20, it was acquired by the government of Nicaragua, and assigned the serial number GN 98. For operations in the United States, it was flown as N74045.

1963–67: It went to Aero Enterprises, La Porte, Indiana, as N6362T. Refurbished at California Aero Sport Aviation, Chino, California, it then went to the Bolivian Air Force on June 10, 1966. It received heavy damage with the Bolivians. The damaged remains were shipped to Cavalier Aircraft Corp. in September 1967.

1968–2016: The aircraft project was stored until acquired by Mark Timken. In 2012, it was purchased by Stewart McMillan. Wings, flaps, and ailerons were built up by Odegaard Wings Inc., Kindred, North Dakota. In 2013, restoration was begun by American Aero Services, New Smyrna, Florida. It was then acquired by the Collings Foundation in 2016. First post-restoration flight was on June 15, 2016. The aircraft was recognized as the Grand Champion Warbird at EAA AirVenture 2016.

Current Owner: Collings Foundation, Stowe, Massachusetts

Right: Seen at California Aero Sport Aviation at Chino, California, in June 1966, preparing for a test flight after the Mustang had been rebuilt for the Bolivian Air Force. It carried the Bolivian Air Force serial FAB 510. (Veronico Collection)

Left: Still on the USAF's inventory, 44-84655 sits at the California ANG base at Van Nuys on October 6, 1957. There is an "O" before the serial number, denoting an obsolete type. The lengthened canopy of the TF-51 conversion is evident in this view. (B. C. Reed via Milo Peltzer)

American Aero Services of New Smyrna Beach, Florida, was engaged to restore the TF-51D for the Collings Foundation's Living History program. When in service with the West Virginia ANG, the aircraft wore the name and artwork of *Toulouse Nuts*. (Roger Cain)

P-51D- 44-84786 (F-6D)

Built: Dallas, Texas; June 1945

Service

1945: Served at various stateside bases including Andrews, Washington, DC; Stuttgart, Arkansas; Brooks Field, Texas; Topeka, Kansas; Hobbs Field, New Mexico; Spokane, Washington; Kelly Field, Texas; and Pope Field, North Carolina.

1949: Flown to McClellan AFB, California, for storage, it was subsequently struck from the USAF's inventory.

1952–81: It was sold surplus and stored until the early 1960s. It was purchased disassembled by Mike Coutches from an aircraft parts dealer in Sacramento, California. In 1966, it was sold as a project to William Myers of St. Charles, Missouri. It was then acquired by Butch Schroeder in 1981, and registered as N5484V.

1981–93: It was sent as a restoration project to Mike Vadeboncour's Midwest Aero Restorations, Danville, Illinois. It emerged in the markings of Clyde B. East's *Lil' Margaret*.

1993: First post-restoration flight on June 17. It won the EAA AirVenture Grand Champion Warbird award.

2017 to present: It was sold to Giovanni Marchi, Ceresara, Mantua, Italy. The aircraft was involved in a take-off accident on May 25, 2018. It was then shipped to Airmotive Specialties, Inc., in Salinas, California, for restoration.

Current Owner: Giovanni Marchi, Museo Volante, Airfield San Martino, Ceresara, Mantua, Italy

Current Scheme Honors:

Capt. Clyde B. East, a 14.5-victory ace, and his F-6D 44-14306, which flew with the 15th Photo Reconnaissance Squadron, 10th Photo Reconnaissance Group, 9th Air Force.

Right: Butch Schroeder taxies *Lil' Margaret* to the Mustang Corral at EAA AirVenture 2005. Notice the camera ports on the aft fuselage. The upper port typically took a K-22 aerial camera, while the lower mount could be fitted with a K-17 or K-24. (Nicholas A. Veronico)

Left: Butch Schroeder found F-6D 44-84786 in a garage in St. Charles, Missouri. "Mustang Mike" Coutches bought this aircraft disassembled from a parts dealer in Sacramento. Coutches was buying P-51H parts, and eventually the parts dealer told him he could not buy any more H parts unless he purchased the F-6D. This fuselage sat in Coutches' backyard for a few years and his kids played in and around it. (Courtesy of Butch Schroeder)

High angle shot of *Lil' Margaret* shows the underwing hard points and three sets per side of zero-length rails for three-inch rockets. (David Leininger)

P-51D-25-NT 45-11439

Built: Dallas, Texas; June 1945

Service

1945–56: Delivered directly to storage at Kelly AFB, Texas. Its first assignment was to the 191st Fighter Squadron, 140th Fighter Group, Utah ANG. The Mustang subsequently served with the Colorado ANG.

1961–93: It was registered to Norton Smith of Seattle, Washington. Smith was killed when the Mustang crashed on May 11, 1961. The wreckage was stored.

1993–2007: The project was purchased by Bill Yoak. It was registered as N51HY and rebuilt, which took 14 years.

2007 to present: It was flown extensively by Bill Yoak, until his passing in 2013. Scott "Scooter" Yoak is now flying *Quick Silver* on the airshow circuit.

Current Owner: Quick Silver Airshows, Las Vegas, Nevada

Current Scheme Honors:
All veterans.

Opposite: The paint scheme of Scott Yoak's *Quick Silver* is quite elegant, with the dark nose, highly polished fuselage, star-and-bar, and North American Aviation logo on the tail. (David Leininger)

Right: Scott Yoak flies an outstanding aerobatic routine in *Quick Silver* and is seen here at EAA AirVenture putting the Mustang through its paces. (David Leininger)

Left: Nose detail of the mirror-like lettering on *Quick Silver*. (Roger Cain)

P-51D-30-NT (TF-51D) 45-11471

Built: Dallas, Texas; July 1945
Service
1945–58: It was delivered directly to storage at Kelly AFB, Texas. Its first assignment was with 27th Fighter Wing (Strategic Air Command), Kearney, Nebraska. The Mustang then served with various ANG units until it was sold at McClellan AFB in May 1958, for US$1,003.45 to Bruce Jochim.
1962–93: It spent time with various owners in the United States including Bruce Jochim (N5418V); Samuel Mallen; the Maytag family, who registered the aircraft as N332; David Zeuschel; and James Barkley, who perished when the aircraft crashed on August 21, 1979. The wreckage was rebuilt incorporating an Israeli Air Force P-51D wing and fuselage. It was rebuilt with radiators in the wings and no scoop as the racer *Stiletto*, and campaigned by Alan Preston and, later, Denny and Scott Sherman.
1984: Skip Holm flew *Stiletto* to victory in the Unlimited Gold race at an average speed of 437.621mph at the National Championship Air Races in Reno, Nevada.
1993 to present: The Mustang was stored disassembled at the Museum of Flying in Santa Monica, California, and was rebuilt to a dual control TF-51D by Peter Regina of Van Nuys, California, incorporating new wings (from Cal Pacific Airmotive, Salinas, California) and dual-control work by Kent Rockwell. It was restored as *DiamondBack*.
Current Owner: Mark Peterson, Mustang High Flight LLC., Boise, Idaho (N51ZM)

Current Scheme Honors:
Aircraft of the 360th Fighter Squadron, 356th Fighter Group, Eighth Air Force at RAF Martlesham Heath.

Left: Its racing career over, *Stiletto* sits in storage at Santa Monica's Museum of Flying, awaiting an uncertain fate. It would soon be put back to stock configuration and emerge as *DiamondBack*. (Nicholas A. Veronico)

Right: *Stiletto* in the pits during the National Championship Air Races at Reno, Nevada. Skip Holm flew this highly modified Mustang to win the Unlimited Gold Championship race in 1984. Notice there is no radiator scoop under the belly, as the radiators have been moved into the wings. (Nicholas A. Veronico)

DiamondBack closes on the camera ship showing the tall vertical tail and lengthened canopy of a dual-control TF-51D Mustang. When it was restored from a racer to TF-51D configuration, fuselage parts from an Israeli Air Force Mustang were incorporated with new-build wings. (Jim Dunn)

P-51D-30-NT 45-11636

Built: Dallas, Texas; February 1945
Service
1945–48: Initially, the Mustang was stored at Kelly Field, Texas. It then moved to 27th Fighter Wing and 82nd Fighter Wing, Strategic Air Command.
1949–56: It saw service with the Colorado, Washington, and South Dakota ANGs.
September 1956: It went into storage at McClellan AFB, California.
1960–66: It was sold to Charles Stark, and then to Frank G. Tallman/Tallmantz Aviation, Santa Ana, California. It was sold at the Tallmantz Auction to Richard Vartanian, then to Steve Roberts, and then John E. Dilley.
1968–2013: It went to Michael W. Bertz, Broomfield, Colorado, who named it *'Stang Evil*.
2014 to present: It was sold to Mark Bingham, Lakewood, Colorado.
Current Owner: Mark Bingham, Flying Tiger Aviation, Lakewood, Colorado.

Current Scheme Honors:
Aircraft flown by the Colorado ANG.

Right: N5467V is seen at the Tallmantz Movieland of the Air at Orange County Airport in the early 1960s, with a Wildcat and a couple of Corsairs in the background. (Doug Fisher Collection)

Left: Mike Bertz brought *'Stang Evil* to Reno for the 1979 air races. The aircraft wore race number 60, applied with tape on both sides of the fuselage. (William T. Larkins)

Mark Bingham flies *'Stang Evil* for an airshow crowd in Colorado. Note that Bingham maintains prior owner Mike Bertz's name on the starboard canopy rail in tribute to the man who cared for this Mustang for 45 years. (Luis Drummond)

F-51D 67-22579

Built: Cavalier Aircraft Corp., Sarasota, Florida; 1967
Service
1967: It went to the Bolivian Air Force, serial FAB-519.
1977–84: It was then acquired by Arny Carnegie and registered as C-GXRG. It next sold to Neil McClain.
1985–2005: It sold to Robert Hester as N52BH, and then to Russell McDonald as N251RM.
2006 to present: It went to John Bagley, Rexburg, Idaho. In 2009, it was reregistered as N551BJ.
Current Owner: John Bagley, Rexburg, Idaho

Current Scheme Honors:
All three of the P-51s Roland Wright flew were named *Mormon Mustang,* two of which were 44-14868 and 44-73219. He was a pilot with the 364th Fighter Squadron, 357th Fighter Group, Eighth Air Force. Wright was credited with three aerial victories, one of which was an Me 262 jet. He rose to the rank of brigadier general with the Utah ANG.

Opposite: John Bagley flies P51D 57-22579 *Mormon Mustang* in company with his P-63A 42-69021. *Mormon Mustang* wears three kill marks attained by 1st Lt. Roland Wright while flying with the 357th Fighter Group in 1945. (Keith Charlot)

Right: John Bagley at the controls of *Mormon Mustang.* The aircraft's tall vertical fin modification has been removed to give the aircraft the stock P-51D appearance. (Keith Charlot)

Left: C-GXRG, seen shortly after its return from Bolivia. Note the Cavalier conversion tall tail and the FAB camouflage paint scheme. (Jerry Liang)

CHAPTER 3
MUSTANG GATE GUARDIANS: IN FROM THE COLD

More than 75 years after the type's first flight, a number of genuine P-51 Mustangs reside as gate guardians. In decades past, these Mustangs were mounted on poles and exposed to the elements. Sitting outside, the fighters quickly deteriorated from the inside out and were often the target of vandals. Thankfully, most of the Mustangs on poles in the United States have been rescued or, at least, lowered to the ground and are now sitting on their gear; others have been rescued and restored.

One of the earliest pole-mounted Mustangs was P-51A 43-6274. This Allison-powered P-51 was accepted by the USAAF on April 26, 1942, and flown to its first assignment at Waycross, Georgia. The following month, it was assigned to Memphis, and then to Nashville to serve with the Fourth Fighter Group. In August 1943, P-51A 43-6274 was redesignated a TP-51A trainer and assigned to the 568th BU, an operational training unit, at Brownsville, Texas. From there, the TP-51 flew from Greenwood, Mississippi, until the end of the war. At the conclusion of hostilities in Europe, TP-51A 43-6274 was flown north and stored at Altus Army Air Field, Oklahoma. At Altus, 43-6274 joined more than 2,600 stored B-17 and B-24 bombers, along with hundreds of tactical fighters ranging from P-38s and P-40s to P-47s and various other models of P-51s.

Transferred under the RFC's schools and memorials program in 1946, 43-6274 ended up in front of a middle school in Frederick, Oklahoma. Had it not become a war memorial in America's heartland, it would have been scrapped with the other aircraft at Altus.

This P-51A was not on the pole very long, as it was purchased to compete in the September 1948 National Air Races in Cleveland, Ohio. Although it arrived at the Cleveland race site with Tommy Mason at the controls, 43-6274 did not complete. By 1950, the Mustang had been sold to Harry McCandless and Ben Widfelt, who kept the aircraft at Council Bluffs, Iowa. Three years later, they sold the fighter to Wally Erickson of Minneapolis, Minnesota, who stored the aircraft for more than 20 years. Erickson later sold the Allison-powered Mustang project to John Crocker, who, in 1978, sold it to Charles Nichols of the Yanks Air Museum at Chino, California.

Once at the Yanks Air Museum, 43-6374 was put into the restoration queue, and its rebirth was guided by the late Stan Hoefler. It took more than a decade to finish, but, when it emerged, Hoefler's attention to detail was obvious, as 43-6374 looked pristine. The Mustang was rebuilt as an F-6A photo-reconnaissance Mustang in the colors of the 67th Reconnaissance Group. All of the photographic equipment has been installed in the aircraft, including the large format, side-facing camera behind the pilot's seat. This restored A-model Mustang is today displayed at the Yanks Air Museum's Chino facility.

D-Model Mustang on a Pole
Another Mustang that sat on a pole, far longer than 43-6274, was Dallas, Texas-built P-51D 44-84900, which was delivered on September 4, 1945. This aircraft flew from the factory to Brookley Army Air Field,

One of the first Mustangs on a pole was P-51A 43-6274, which was parked at the storage depot at Altus Army Air Field, Oklahoma. In 1946, the Allison-powered fighter became a memorial at the Frederick Middle School, Frederick, Oklahoma. It is interesting to note that the school's main building, with the pointed roofs, is still present today, while the domed church in the background has been rebuilt, and the Mustang's former roost is a field with construction now taking place. (Martin Kyburz/Swiss Mustangs collection)

Alabama, before it was requisitioned with some other D models by the National Advisory Committee for Aeronautics (NACA, the predecessor of today's National Aeronautics and Space Administration, NASA) and flown to its facility at Langley, Virginia. P-51D 44-84900 lost its military identity and became "NACA 127" while flying as an airborne testing laboratory.

NACA's most obvious external modification to the Mustang is its tall vertical tail, which increases lateral stability during high-speed dives. The aeronautics agency also fitted the top of the outer wing panels with model test sections. The test sections are painted white and located outboard of the wing gun bays.

"Pilots would notice during high-moisture days that they could see the flow along the upper surface of the wings, during high-speed pull-ups," said William C. "Bill" Allmon, owner of NACA 127. "They determined that if they took the plane up to a certain altitude, did a high-speed dive and a 4g pull-up, they could get the flow over the wing into transonic speeds. It was basically a flying wind tunnel."

NACA Langley researcher Robert R. Gilruth, then-chief of the Flight Research Section, developed the "wing-flow method" of testing in 1944, and a number of NACA-operated Mustangs were outfitted for this special mission. Some had cameras that recorded the airflow over and around the airfoils, while NACA 127 had cameras mounted inside the fuselage and wings to photograph a variety of instruments reading data from strain gauges. The strain gauges were measuring pressures, as transonic airflow passed over the models mounted on top of the wing. Some models were airfoil shapes, while others were semi-span aircraft (essentially half an aircraft split lengthwise and mounted flush with the testing surface).

On each wing, outboard of the test model is a small square airfoil that protrudes into the airstream. Both airfoils provide yaw indications to an instrument, similar to a pilot's directional indicator (PDI), to tell the pilot when the aircraft is being flown perfectly straight. Only when there was no yawing to either side could the test begin. This indicated to the pilot and engineers that the test was performed correctly, and that the data collected was valid.

Typically, a test pilot would take the aircraft and its experiment up to 35,000ft and then push over into a dive reaching Mach .71 to .73. Descending to 10,000ft, the pilot would begin a 4g pull-out while recording data. Then the pilot would climb up to 35,000ft and perform the test again, and again, and again. At Mach .71 to .73 in a 4g pull-out, the airflow over the wing model would, locally, go supersonic. Capturing how the airflow reacted as it passed around the airfoil model and how it accelerated from subsonic to supersonic flow was, at the time, extremely difficult to model in a wind tunnel. Wind tunnel testing, as well as "wing-flow," "drop-body" (airfoil and aircraft shapes mounted on the nose of a bomb dropped from a B-29 at 30,000ft), and "rocket model" (air foil shapes mounted to the nose of a rocket) tests, combined with captured German aeronautical information provided the data used to design America's postwar generation of jet aircraft.

In 1952, 44-84900 was released from NACA service and acquired by the Pennsylvania ANG. The guard used the aircraft as a fighter for a number of years, having reinstalled much of the armament. The Mustang was then grounded for use as an instructional trainer. Many Pennsylvania ANG mechanics in the mid-1950s honed their skills on this tall-tail Mustang. In 1973, it was mounted on a pole and displayed

Above and overleaf: Wally Erickson had bought the early, high-back Mustang in the 1950s and kept it stored until John Crocker was able to make a deal for the A model. Charles Nichols of the Yanks Air Museum, Chino, California, acquired the Allison-powered Mustang project from Crocker. The project's fuselage arrived on a flat-bed truck in 1978. (Yanks Air Museum)

Mustang Gate Guardians: In from the Cold

at the Greater Pittsburgh ANG Base as a gate guardian. Over the years, 44-84900 was displayed in an overall bare metal finish and was later painted silver with ANG markings. In the late 1980s, while still on the pole, the Mustang was given a World War Two combat paint scheme with fuselage codes AJ*S.

While the Mustang was on the pole, the rumor got started, and stuck, that 44-84900 was "the only P-51 to ever land on an aircraft carrier." A Mustang did, in fact, land on an aircraft carrier; however, P-51D 44-8490 was built too late to participate in those trial flights. The testing was known as "Project Seahorse," and Lt. Robert M. "Bob" Elder flew P-51D 44-14017, redesignated an ETF-51D, for this program. Although NACA 127 was also designated an ETF-51D, 44-84900 never went aboard ship.

Back in the Air, But Not on a Pole

Bill Allmon started flying gliders as a young man, and he raced them for a long time. Using a private aircraft in his business, he transitioned into more complex aircraft as his ratings evolved. Allmon purchased a Beech Starship as his business grew, but for personal pleasure he had always wanted a P-51 Mustang. Allmon said:

> I had decided I wanted a Mustang and looking around the market, any of those nice ones were starting to top $1 million at the time. I couldn't afford that. I could, however, afford to cash flow a restoration. That's kind of the way I started. I contacted Mark Clark [of Courtesy Aircraft Sales], as he represented a couple of airplanes and they were too expensive, but this project intrigued me because he said it actually had a history.

Opposite: After a decade-long restoration, 43-6274 emerged as an F-6A photo reconnaissance Mustang wearing the colors of the 67th Reconnaissance Group. Notice the camera looking through the rear window of the cockpit. (Nicholas A. Veronico)

Allmon went back east in September 1993 and saw the project Mustang at the Air Heritage Inc. museum in Beaver Falls, Pennsylvania. "It was pretty rough looking when I first saw it," Allmon said. The Mustang project sat in the corner of the museum with the wings sitting side-by-side, the tail was off, and the canopy was slid all the way back.

David Tallichet had made a deal with the Air Force to acquire the plane, trading them two Fiberglas replica aircraft in exchange for the Mustang. Tallichet never registered the project and had not done any work to it when his restaurant empire suffered some financial setbacks, and he was forced to sell off part of his massive 120-plus aircraft warbird collection. Allmon was interested in the project and a deal was made. After arranging payment, Allmon received clear title to the Mustang from the Air Force. In March 1994, the P-51D project was trucked to John Muszala's Pacific Fighters at Chino Airport, California. Restoration work got underway, and then the project was moved to Idaho Falls in 1996, when Muszala relocated his shop. Allmon recalled:

> The thing that did the most damage to the airplane was just sitting. It sat out in the weather for a long time. Probably thirty-plus years. Certainly from the lower longeron down it was a badly corroded mess. I think that's what motivated the Air National Guard to get rid of it.

Mustang Gate Guardians: In from the Cold

Factory-fresh P-51D 44-84900 as delivered to NACA's Langley Memorial Aeronautics Laboratory in 1945. The aircraft has not yet been modified for NACA service. The Mustang wore NACA serial number 127 during its testing days. (NASA)

All the rot was happening down below, and it was starting to affect the structural integrity of the aircraft where it was attached to the pole. The higher up you went in the airplane, the better things were, but all that low stuff, the lower longerons were like a croissant. They were forged in layers, and you could actually peel the metal away. The scoop of the doghouse had to be totally redone. The belly skin was gone. The intake trunk was gone; it was corroded, totally eaten away. When we opened it up it, we found that every creature that could get into that airplane had lived in it. It was a nest for everything.

The Mustang sat on the pole with its original engine, but, at some point, the propeller hub had been changed and the blades were swapped out. Mustang propellers are expensive, and one had to be tracked down.

The restoration started out as a rebuild to combat configuration, but as Muszala's team got deeper and deeper into the aircraft, they kept finding the unique fittings of the NACA modifications. This aircraft had never been civilianized, and the decision was made to switch course 180 degrees. All of the recently installed armament was removed, and the Mustang was brought back to NACA configuration. This was one of the standard-setting Mustang restorations, as Muszala's team photographed nearly every part that was taken off, both externally and internally. Allmon explained:

In doing that, we found the inside of the skin was marked with doodles and messages and check marks or sign-offs, not just inspection stamps, done by employees at the factory. Muszala's team took great pains to put all that back the way it was when we found it. We photographed each one of them and carefully recreated them. Now, they do it a lot. Midwest Aero does it and a few other people do it, but this was the first airplane that went through such a detailed restoration process.

A NACA technician checks the strain gauges inside the wing that will monitor airflow over the semi-span model mounted on the Mustang's wing test section. That model bears a strong resemblance to the Bell X-1. (NASA)

When the Mustang was originally built, all of the parts were painted before they were assembled, so it was totally bare-riveted throughout. Allmon continued:

> When we were taking it apart, we were meticulous at putting those colors back on each rib – even though it's a pain in the ass to do – just so it's period perfect. It really looks like it came off a factory line. One good example is the engine mount. When the cowls are removed, the engine mount is in four different colors, but that's how it came from the factory. The airplane was restored exactly the way it came apart and that's how it went back together.
>
> During the restoration, we had tried real hard to get information out of NASA and initially they were pretty closed-mouth about things. They didn't supply much, but, as the restoration went along and, I guess, at some point they started to feel like I was serious about preserving it as a test airplane, they became a lot more forthcoming. I was working with a gentleman out of Virginia, and he got in touch with one of the engineers. One thing led to another, and the engineer volunteered a wing flow model, one of the actual test articles. His experiment is the airfoil you see on the left wing. He gave me his test pictures as well.
>
> I later met NACA test pilot Bob Champine who flew this aircraft. In talking to Bob, they didn't fly the airplane much. He said that once a mission was flown and the experiments they needed were verified as having been done correctly, they just parked it until the next mission. When another experiment was planned, they would then go out and prepare the aircraft, so they didn't accumulate many flight hours.

Continuing the NACA/NASA connection, during the restoration, 44-84900's data plate was flown aboard the Space Shuttle *Discovery* on STS-85, a mission that lasted from August 7–19, 1997. STS-85 was the Space Shuttle program's 86th flight, and *Discovery* cruised above the Earth at an altitude of 168 miles and reached a maximum speed of 17,610mph on that mission. That makes 44-84900's data plate the fastest Mustang data plate – ever!

Muszala's team worked on the Mustang for three years, and had it ready to fly in July 1998, just a few weeks before EAA AirVenture at Oshkosh, Wisconsin. There, it won the Post-World War II Grand Champion award and Muszala's Pacific Fighters took home the coveted Golden Wrench Award.

In September 2010, 44-84900 was entered in the National Aviation Heritage Invitational Trophy competition, which is held in conjunction with the National Championship Air Races at Reno, Nevada. Here the NACA Mustang attracted a great deal of attention and garnered top honors, receiving the Rolls-Royce Aviation Heritage trophy. This perpetual trophy resides at the Smithsonian Institute's National Air and Space Museum Steven F. Udvar-Hazy Center in Chantilly, Virginia, and Allmon's name is engraved as the 2010 winner. In September 2013, Allmon, Muszala, and the NACA Mustang were invited back to the National Aviation Heritage Invitational as a returning champion, where the unique Mustang was a crowd favorite. Allmon said:

> I've put about 350 to 360 hours on it. And recently I've taken some of the NACA equipment out of the plane so I can put a bag or two in it, and we've put the back seat in it. We can always change it back to the full NACA configuration pretty easily. It takes a couple days to do, but we made it easy to pull all that standard stuff out and put all the NACA stuff back in.

The cockpit is fairly stock, with modern radios hidden behind one of the NACA panels. A set of Collins Pro-line radios and a transponder are grouped onto a plate that can be swapped with the original armament panel low in the cockpit behind the control stick. Using butterfly nuts, the plate with the radio heads and the original armament panel can be changed out in a matter of minutes.

Although 44-84900 sat on a pole for more than 30 years, being there kept the aircraft from being civilianized. The internal modifications found during the restoration enabled the aircraft to be brought back to the days when NACA used diving aircraft to replicate airflow over wing models at transonic speeds. P-51D-25-NT 44-84900 is a true time capsule of aviation history.

The Mystery P-51H Project
Hayward, California, aircraft dealer Mike Coutches (1922–2016) was known as "Mustang Mike." In the mid- to late-1950s, Coutches would acquire surplus P-51 Mustangs and Mustang components from sales at McClellan AFB, near Sacramento. Most of the Mustangs he acquired were low-time examples. He would fly them to Hayward, inspect and repair as necessary, then sell them for US$3,995, licensed and ready to fly. His business handled more than 20 P-51s, and that is how he earned the nickname.

Located across the Hayward Airport from Mustang Mike's business was the 144th Fighter Group, 194th Fighter Squadron of the California ANG. The unit flew lightweight P-51H model Mustangs, and Coutches fell in love with that variant. He was able to acquire P-51H-5-NA, serial number 44-64314, N551H – one of only six H models known today, and the only one currently flying.

P-51D 44-84900 with the NACA tall-tail modification to reduce yaw. The aircraft is in service with the Pennsylvania ANG. Note that all of the external NACA test gear has been removed and the fuselage bears the remnants of a buzz number under the cockpit. (via A. Kevin Grantham)

Above: A little later in its ANG career, 44-84900's markings have been updated. Notice the "0" preceding the serial number denoting an aircraft that is more than ten years old. (Bill Allmon Collection)

Opposite: P-51D 44-84900, albeit missing the last digit in this paint scheme, was retired to a pole to serve as a gate guardian at the Greater Pittsburgh ANG Base. Years of sitting outside in the harsh Pennsylvania winters took their toll on the Mustang, and it was removed from the pole when the aircraft's structural integrity became an issue. (via Earl Holmquist)

Left: The restored cockpit of 44-84900 includes all of the testing equipment when flown in NACA service. (Pacific Fighters)

Opposite: Mustang owner Bill Allmon (seated) and John Muszala (standing) and his Pacific Fighters of Idaho Falls, Idaho, collaborated to return 44-84900 to its NACA configuration. Allmon did a lot of research and interviewing many of the ex-NACA pilots, finding original reports and learning a great deal about the aircraft's testing days. The original NACA equipment was refurbished or replaced for this restoration. (Pacific Fighters)

In conversations with Coutches, he had mentioned that he had another complete H model under restoration, but no one had ever seen it. It lived for years as a rumor within the warbird community, and while the rumor circled, Coutches was scouring the country for surplus H model parts – parts that no one had a use for, as there were so few P-51H models that needed them.

Coutches was getting up in age by the second decade of the new millennium, and simultaneously,

the City of Hayward was forcing his business, American Aircraft Sales, to make extensive and expensive modifications to its hangar and offices. Coutches' son Robert, who now ran the business, decided it was more economical to move the business less than 20 miles east to the Livermore Airport, where they could acquire a new, larger, more modern facility. The move would also give the family the opportunity to consolidate its many aircraft in one hangar.

One of the first to move was the family's super rare, low-time Grumman F8F-1 Bearcat, Bureau number 121679, N818F. Shortly thereafter, another aircraft needed to be moved and a truck and trailer from Taigh Ramey's Vintage Aircraft in Stockton, California, was called to the Coutches home in Fremont. From Coutches' shop behind the garage rolled the fuselage of a P-51H that was undergoing restoration. With the fuselage were a set of wings, still in the crate from when they were packed in the mid-1960s. Mustang Mike had acquired P-51H-5-NA 44-64203, which had been built in Inglewood in June 1945. After the war, 44-64203 flew with the New York ANG, and when its career was over, the fighter became a gate guardian at Sampson AFB, New York. When the base closed, 44-64203 became a memorial in front of the American Legion Post at Geneva, New York. By the time Mustang Mike had acquired this P-51H, it had sat out in the weather for years, been subject to vandals, and its main gear was sunk in the mud to its axels. If Coutches had not rescued 44-64203, it surely would have been scrapped as an eye sore.

So, as it turns out, the P-51H rumor was fact, and Mustang Mike did indeed own two of his favorite type. His son Robert is planning to inventory the massive parts cache and resume the restoration in early 2023.

Above: The wing flow model on the left wing of NACA 127's test section is a genuine test shape from the late 1940s. This attention to detail makes this Mustang a crowd favorite. (Nicholas A. Veronico)

Right: NACA 127's Bill Allmon (right) receives the Rolls-Royce Aviation Heritage Trophy at the National Aviation Heritage Invitational competition at Reno, Nevada, in September 2010. The competition is held in conjunction with the National Championship Air Races. The trophy has since been renamed in honor of astronaut Neil A. Armstrong, and is displayed at the Smithsonian Institute's National Air and Space Museum's Steven F. Udvar-Hazy Center in Chantilly, Virginia. (Nicholas A. Veronico)

John Muszala II tucks the NACA Mustang in close over Nevada's Pyramid Lake. Known as NACA 127 when in service as an aeronautic research tool, this Mustang served as a high-speed flying testbed. Results from these tests, coupled with what was learned in wind tunnel tests along with captured German swept-wing research, led to the jet-powered fighters of the late 1940s and 1950s. (Richard H. VanderMeulen)

"Mustang Mike" Coutches wanted a P-51H and spent years gathering H model parts. Coutches built up this beautiful example, 44-64551, which is now being flown by his son Steve. (Nicholas A. Veronico)

Seen outside the American Legion Post at Geneva, New York, is ex-ANG P-51H-5-NA 44-64203. The aircraft served in Alaska before being transferred to the US East Coast. By the time Mike Coutches saw the aircraft in 1965, the spinner was gone, the canopy had a chunk missing, and the wheels had sunk up to their ankles. The best part about the aircraft was that it is a time capsule of a late World War Two Mustang. (Rick Turner Collection)

The fuselage of P-51H 44-64203 rolls out of Mike Coutches' shop and is ready for transport to American Aircraft Sales' facility at the Livermore Airport. Most in the warbird community had heard the rumors that another complete P-51H existed, but no one had seen it for 50 years. (Taigh Ramey)

P-51D-25-NA 44-72948 has sat outside since 1957 and wears the nose art of *Wham Bang* on the lower port side cowling. The aircraft has been removed from the pylons and now sits on its landing gear. (Veronico Collection)

Wisconsin's Volk Field is home to P-51D 44-72989, which was delivered in February 1945. Its last posting was with the Wisconsin ANG, and it is now displayed at the entrance to the Volk Field ANG Base near the town of Camp Douglas, Wisconsin. (Jerry Liang)

There are two P-51Ds displayed outside in South Korea. P-51D-25-NA 44-73494 sat at the gate at Yongdungpo Air Base in Seoul and is seen there in 1967. This aircraft was moved to the Korea War Memorial in Seoul in 1995. The second aircraft is believed to be a TEMCO TF-51 conversion, serial number 44-84669 (possibly ROKAF 51-8424). (Robert F. Dorr)

Above, opposite and overleaf: P-51D 44-73972 is one of the lucky Mustangs, having been brought down from its pole. The rare fighter is now on its gear at the California ANG's base at the Fresno Air Terminal in central California. (William T. Larkins/Nicholas A. Veronico)

Subsequent to this photo, P-51D-30-NA 44-74407 was removed from its pole mounting and now resides within the 122nd Air Wing's Heritage Air Park at Fargo International Airport, North Dakota. The North Dakota ANG flew the Mustang from 1947 to 1954. (Veronico Collection)

Mustang Gate Guardians

P-51D 44-72948	West Virginia ANG, Charleston
P-51D 44-72989	Volk Field, Wisconsin
P-51D 44-73494	Korean War Museum, Seoul, South Korea
P-51D 44-73972	California ANG, Fresno
P-51D 44-74407	North Dakota ANG, Hector Field
P-51D 44-72123	San Isidro, Dominican Republic (ex-55th FG)
P-51D F-303*	Jakarta-Halim Perdanakusuma International Airport
P-51D F-347*	Armed Forces Museum, Jakarta, Indonesia
P-51D F-363*	Malang AFB, East Java
P-51D FAG-336*	Guatemalan Air Force, La Aurora Air Base

* = USAAF serial number unknown

In total, the Guatemalan Air Force (Fuerza Aerea Guatemalteca, FAG) flew thirty P-51Ds. Don Hull of Sugarland, Texas, returned six of the FAG P-51s to the US civil market, and part of the deal was that Hull would take existing parts and build a Mustang Memorial. Given the serial FAG 336, the Mustang is well maintained in this attractive roadside memorial. (Tony Veronico)